ALTERNATIVE FORMULATIONS AND PACKAGING TO REDUCE USE OF CHLOROFLUOROCARBONS (CFCs)

ALTERNATIVE FORMULATIONS
AND PACKAGING
TO REDUCE USE OF
CHLOROFLUOROCARBONS
(CFCs)

by

Thomas P. Nelson and Sharon L. Wevill

Radian Corporation
Austin, Texas

NOYES DATA CORPORATION
Park Ridge, New Jersey, U.S.A.

Copyright © 1990 by Noyes Data Corporation
Library of Congress Catalog Card Number: 90-7746
ISBN: 0-8155-1257-0
ISSN: 0090-516X
Printed in the United States

Published in the United States of America by
Noyes Data Corporation
Mill Road, Park Ridge, New Jersey 07656

10 9 8 7 6 5 4 3 2 1

Library of Congress Cataloging-in-Publication Data

Nelson, T.P.
    Alternative formulations and packaging to reduce use of
chlorofluorocarbons (CFCs) / by Thomas P. Nelson and Sharon L.
Wevill.
        p.   cm. -- (Pollution technology review, ISSN 0090-516X ; no.
194)
    Includes bibliographical references and index.
    ISBN 0-8155-1257-0 :
    1. Pressure packaging. 2. Aerosol propellants.
3. Chlorofluorocarbons. I. Wevill, Sharon L. II. Title.
III. Series.
TS198.P7MN45   1990
688.8--dc20   90-7746
                                        CIP

Foreword

This book describes alternative formulations and packaging techniques for the reduction or elimination of chlorofluorocarbon (CFC) use as an aerosol propellant. Use of CFCs in specific categories of aerosols considered "nonessential" was banned by the U.S. in 1978. Recent renewed interest in further reducing worldwide production and consumption of CFCs, and other chemicals implicated in the depletion of the earth's stratospheric ozone layer, is responsible for this study, which covers currently exempted and excluded CFC aerosol applications and their alternatives.

The book is presented in two parts. Part I gives background information on the issue and an overview of technically feasible methods for reducing CFCs in aerosol products without adverse effects on human life and health, military preparedness, and the economy. Part II discusses industry's experience in converting to alternative formulations. Detailed non-CFC formulations are provided for 28 categories of aerosol products. Special equipment may be needed to include these formulations in aerosol containers, and this is discussed along with a variety of alternative dispensing devices. Advantages and drawbacks of these devices are discussed in detail, and examples of consumer products which have successfully utilized these alternatives are given.

The information in the book is from:

Alternative Formulations to Reduce CFC Use in U.S. Exempted and Excluded Aerosol Products, prepared by Thomas P. Nelson and Sharon L. Wevill of Radian Corporation for the U.S. Environmental Protection Agency, November 1989.

Aerosol Industry Success in Reducing CFC Propellant Usage, prepared by Thomas P. Nelson and Sharon L. Wevill of Radian Corporation for the U.S. Environmental Protection Agency, November 1989.

The table of contents is organized in such a way as to serve as a subject index and provides easy access to the information contained in the book.

NOTICE

Contents and Subject Index

PART II
ALTERNATIVE FORMULATIONS AND
AEROSOL DISPENSING SYSTEMS

Part I

Background and Overview

The information in Part I is from *Alternative Formulations to Reduce CFC Use in U.S. Exempted and Excluded Aerosol Products,* prepared by Thomas P. Nelson and Sharon L. Wevill of Radian Corporation for the U.S. Environmental Protection Agency, November 1989.

1. Introduction

The use of chlorofluorocarbons (CFCs) in specific categories of aerosol propellant use considered "nonessential" was banned in the U.S. by regulations promulgated in 1978 (1). An aerosol was defined as a package comprising a self-pressurized, non-returnable container constructed of metal, glass, or plastic that contains a fluid product and that is fitted with a valve for expelling the product as a spray, liquid, gas, foam, powder, or paste. The banned CFC propellants included the fully-halogenated types: CFC-11, CFC-12, CFC-113, CFC-114, and CFC-115.

In view of the recent renewed interest in reducing worldwide production and consumption of CFCs and other chemicals implicated in the depletion of the Earth's stratospheric ozone layer, the U.S. Environmental Protection Agency (EPA) undertook this study of currently exempted and excluded CFC aerosol applications and their alternatives.

The EPA and the Food and Drug Administration (FDA) initiated and supervised the three-stage program from 1977 to 1978 to eliminate most uses of fully halogenated CFC propellants for aerosol propellant applications except for "essential uses." If the CFC in the product functioned as more than a propellant or dispersant, it was considered "the product" or "part of the product" and excluded. The agencies held the concepts of "product" and "propellant" to be mutually exclusive. Also, the EPA and FDA evaluated the need for certain "essential" aerosol propellants. These were products that, for reasons of safety, health, or national security, required a CFC propellant. The initial EPA evaluation resulted in a list of approximately 14 general applications considered exempt.

2

In the years following the transition, a few additional petitions for exemptions were tendered. Some were accepted. In addition, agency clarification of the exclusion of specific product types was requested. Some products were added to the exempt list as the result of these activities. Exemptions were applied to the product category, rather than to specific products or brand names.

This report identifies technically feasible methods for reducing CFCs in aerosol products without adverse effects on human life and health (inhalants, pharmaceutical tablet press lubricants), military preparedness (lubricants for electronic gear), and the economy (cleaners and chill-testers for computer equipment).

MONTREAL PROTOCOL REQUIREMENTS

The August 1988 EPA regulations (2) implement the Montreal Protocol (3). The regulatory mechanism to implement the Protocol is different from the 1978 aerosol propellant ban. Rather than develop regulations specific to each industry application (such as aerosol production), the entire supply of fully-halogenated chlorofluorocarbons will be reduced.

The Montreal Protocol of 1987 and the corresponding EPA Final Rule to implement the Protocol require the following reductions in calculated levels of controlled CFCs:

- Beginning July 1, 1989, a freeze at 1986 consumption and production levels of CFC-11, -12, -113, -114, and -115 on the basis of their relative ozone depletion weights;

- Beginning mid-1993, a reduction of these CFCs to 80% of 1986 levels; and

- Beginning mid-1998, a reduction of these CFCs to 50% of 1986 levels.

In May 1989, delegates of the nations party to the Protocol met in Helsinki and agreed on a five-point declaration to strengthen the Protocol. The five points are as follows:

- **Phase out consumption and production of ozone-depleting CFCs** "not later than the year 2000."

- **As soon as feasible, phase out halons and control** and reduce other **ozone-depleting substances** that contribute significantly to ozone depletion.

- **Accelerate development of environmentally acceptable alternative substituting chemicals, products,** and technologies.

- **Help developing countries by providing information,** training, and possibly funding to facilitate adoption of acceptable alternatives.

- **Urge all states that have not already done so** to join the Protocol.

OBJECTIVES AND ORGANIZATION OF THE REPORT

The purpose of the present report is 1) after reviewing the application, availability, and cost-effectiveness of CFC alternatives, to determine if suitable non-CFC alternatives or alternatives with lower CFC content can be substituted for those aerosol products in the U.S. that still use CFCs, and 2) to determine the steps necessary to convert to the best alternatives.

Section 2 of this report describes exempted and excluded CFC aerosol applications in the U.S., as well as those applications not covered by the regulation.   Section 3 discusses the current U.S. consumption of exempted, excluded, and nonregulated CFC products.   Suggested alternative aerosol formulations are discussed in Section 4, and the economics of their development and application are explored in Section 5.

Section 6 presents the conclusions of the study and discusses the most promising alternative formulations for CFC aerosols remaining in use in the U.S. Based on suggested scenarios for introducing the alternatives, the following reductions in CFC aerosol usage in the U.S. are judged possible (see Sections 4 and 6):

CFC Products	Present Consumption	1990 Usage Scenario One	1990 Usage Scenario Two	1995 Usage Scenario Three	2000 Usage Scenario Four
Mold Releases	1.50	1.50	1.11	0.00	0.00
Lubricants- (electric/electronic)	1.90	1.90	1.10	0.00	0.00
Lubricants-Tablets	1.00	1.00	0.68	0.00	0.00
Solvents- (electric/electronic)	6.00	6.00	4.20	0.00	0.00
MDID (Metered-Dose Inhalant Drugs)	3.90	4.00	4.00	5.25	0.50
Contraceptives	0.10	0.10	0.10	0.00	0.00
Solvents-Medical	0.60	0.60	0.46	0.00	0.00
ALL OTHERS	10.50	0.00	0.00	0.00	0.00
	25.50	15.00	11.65	5.25	0.50

Units: MM lbs/yr.[*]

In addition, CFC control measures may need to be examined for certain self-pressurized products not covered in this report. For example, in 1977, EPA granted an exemption to flying insect sprays used in tobacco barns. The exemption was not limited by container size. Depending on the barn size and layout, one insecticide averaging 35% Vapona (2,2-dichlorovinyl dimethyl phosphate and related compounds) and 65% CFC-12 was used in sizes ranging from 12 Av.oz. to 25 lbs. For this product, considering density, the "aerosol" containers are limited to net weights of approximately 24 Av.oz. (1 lb. 8 oz.) or less.

---

[*]Readers more familiar with metric units may use the conversion factors in Appendix C.

2. CFC Aerosol Applications Exempted in the U.S.

Table 1 lists CFC aerosol applications exempted from EPA's and FDA's 1978 regulations.

EPA's final rule-making excluded aerosols containing propellant only, i.e., aerosols that do not propel another material. In such instances the propellant becomes the product. Also, when the CFC is an active ingredient, the aerosol product assumes a nonregulated status. The concepts of "product, active ingredient, and concentrate" and "propellant" are held to be mutually exclusive, and regulations only deal with aerosol propellants. Table 2 lists the CFC aerosol applications that are excluded from the regulation.

The rationale applied in 1978 and 1979 by EPA and the FDA when considering proposed exemptions included the following:

- The need for a nonflammable product;

- The five years often taken by the Drug Division of the FDA to approve an Amended New Drug Application (applies to bronchodilators and other inhalants, and vaginal contraceptive foams for human use);

- Required solvency and purity profiles, e.g., CFC-113;

- Doctrine of equivalency--that highly similar products cannot reasonably be treated differently;

TABLE 1. SPECIFIC EXEMPTED AEROSOL PROPELLANT APPLICATIONS - 1978

1. Release agents for molds used to produce plastic and elastomeric materials;

2. Lubricants for rotary-type press-punches for the production of pharmaceutical tablets;

3. Lubricants, cleaner-solvents, dusters or coatings for industrial/institutional applications to electronic or electrical equipment;

4. Mercaptan stench warning devices for mines;

5. Other warning devices, such as intruder alarms, boat horns, bicycle horns, etc.;

6. Flying insect pesticides for use in commercial food handling areas, except when applied by total release or metered valve aerosol devices;

7. Propellants for flying insect pesticides for the fumigation of aircraft;

8. Flying insect spray for tobacco barns;

9. Metered dose inhalant drugs, as follows:

    -- Steroid drugs for humans, (oral and nasal)
    -- Ergotamine tartrate drugs, and
    -- Adrenergic bronchodilator drugs (oral);

10. Contraceptive vaginal foams for human use;

11. Aerosols for the maintenance and operation of aircraft;

12. Aerosols necessary for the military preparedness of the United States of America;

13. Diamond grit sprays; and

14. CFC-115 ($CClF_2$-CF_3) for the aeration of puffed food products.

TABLE 2.   EXCLUDED CFC AEROSOL APPLICATIONS

CFC as Active or Sole Ingredient

1.   CFC-12 used as a polyurethane blowing agent (insulation foams);
2.   CFC-12 and CFC-114 mixtures used in tire inflators;
3.   Certain specialty foams, whips, and puffs
4.   Medical solvents such as silicone-based bandage adhesive (CFC-113) and bandage adhesive remover (CFC-113, with 5% CO_2);
5.   CFC-12 and CFC-114 refrigeration and air-conditioning system refill units*;
6.   Drain openers;
7.   Microscope slide cleaners;
8.   Computer cleaners and dusters (equivalent to number 3 in Table 1);
9.   Boat horns;
10.   Halon fire extinguishers (the types containing 95% Halon-1211 and 5% CO_2 may not be excluded);
11.   Intruder alarm sonic devices for homes and cars; and
12.   Skin chillers--for medical purposes.

*The aerosol industry recognizes these products as "aerosols" in surveys and close involvement with the marketing firms.

- Limited availability of substitute propellants or products;

- Stratospheric ozone impact, e.g., production tonnage per year; and

- Life-saving potential of the proposed exempted product, e.g., medical, military, flammable gas alarm systems, etc.

RATIONALE FOR EXEMPTED USES OF CFC AEROSOLS

This section will specifically examine the rationale for exempting certain aerosol products from the 1978 ban on CFC use. Where possible, products in similar groupings will be treated as a unit.

Release Agents (For molds to produce plastics and elastomers for medical applications. Also, for mold chambers in punch presses used to produce pharmaceutical pills and tablets.)

CFC-11 and CFC-12 are used, sometimes with CFC-113, in an industrial environment. The solvent propellant system must be extremely pure to prevent product contamination. The system must also be compatible with the item being formed to minimize such effects as colorant bleed, distortion, and filler micro-crystallization. Since the solvent/propellant system may be used repeatedly, and usually in a controlled environment with negative-pressure settings, minimum flammability, preferably nonflammability, is an important attribute. The solvent must be volatile so that all but ppb traces are gone by the time the product is packaged and possibly radioisotopically sterilized. CFC-11 and CFC-113 provide this volatility, while also modifying the particle size distribution of the spray to achieve optimum surface coating efficacy.

When the application was made for CFC-11, CFC-12, and CFC-113 to be exempted from the regulation, the only available nonflammable propellant was HCFC-22. The possible nonflammable chlorinated solvent alternatives were methylene chloride and 1,1,1-trichloroethane (4). Both contain considerable amounts of mixed inhibitors (free-radical chain stoppers), such as 1,4-dioxane and nitromethane. The effect of these high-solvent substances on molded medical products was sometimes deleterious; they sometimes volatilized from them with difficulty, and the contamination effects of the various inhibitors would have caused the FDA to require a penetrating evaluation of product purity and toxicology.

Lubricants, Cleaner-Solvents, Dusters and/or Coatings for Electronics

Extremely pure CFC-12 and CFC-113 are used to make sprays of varying particle size distribution for treating finished electric or electronic circuit boards, tuners, computer lay-outs, and similar "hi-tech" articles. Where lubricant application or solvent-based cleaning is required, a mixture of about 25% CFC-12 and 75% CFC-113 is the choice. Coating sprays (other than lubricants--such as anti-static) may use 30-35% CFC-12; the remainder is CFC-113. Duster sprays often use 100% CFC-12. Vapor-phase defluxers use various blends of CFC-12/113, depending on the specific application.

CFC-113 is preferred because of its extreme purity when purchased in grades specific for the purpose and because of its compatibility with materials of construction. It can effectively remove oils, solder by-products, greases, inorganic dusts, and other detritus without adversely affecting elastomers, plastics, and metals. Any more critical treatment of the subject would have to consider individual applications.

The firms seeking an exemption claimed that there was no substitute chemical with the extreme purity, nonflammability, materials compatibility, and the volatility of CFC-113. One manufacturer of personal computer boards cited a room where 10 to 15 operators sprayed residue materials from finished boards at the rate of two to five boards per minute. Besides leaving traces of petroleum oils on the boards, the use of hydrocarbon gases, such as iso-butane, would have caused a fire hazard in an area where testing equipment was needed that could not be purchased in explosion-proof designs.

Mercaptan (Thiol) Stench-Type Warning Devices

Typically, these devices are used in mine tunnels to warn of the build-up of seeping methane and/or carbon monoxide flammable gases. The devices are operated by a small, portable sniffer-tester that continuously monitors the air and detects flammable gases by the differentiating action of a Wheatstone Bridge circuit. The stench device is preferred to a siren, since some miners may be operating air hammers or drills and may also have hearing deficiencies because of their long exposure to the 110-decibel noise of these tools.

The typical aerosol container, equipped with a piston-operated actu-ational device, contains an ethyl mercaptan (ethylthiol) stenching agent in a mixture such as one of the following:

3%	Ethyl Mercaptan
17%	1,1,1-Trichloroethane (4)
80%	CFC-12
	or
2%	Ethyl Mercaptan
98%	CFC-12

Mining companies seeking the exemption reported the life-saving attribute of the product, also pointing out that a nonflammable composition was necessary, since 1) the entire can was discharged quite rapidly and 2) there was already a buildup of flammable gases; deliberately releasing more would serve to exacerbate an already hazardous situation.

Other Warning Devices

Other self-pressurized warning devices include home and car intruder alarms, boat horns, etc. They generally consist of 100% CFC-12. Since large quantities of gas are released at one time--often the entire can contents-- under a variety of closed or open environmental conditions, the use of a flammable propellant seemed unwise. (At the time the exemption was sought, no toxicologically safe, liquid nonflammable alternative propellant had been identified.)

The firms seeking an exemption pointed out that products designed for the preservation of life and property should not, of themselves, be hazardous to use. Small (though potentially dangerous) amounts were released each time, and there were no available substitutes. Also, the production volume was fairly low.

Flying Insect Pesticides (For use in commercial food-handling areas, except for total-release and meter-sprays; also for killing insects in and on aircraft and in tobacco warehouses.)

The products for commercial food-handling areas have the following typical formula:

2%	Pyrethrins & Piperonyl Butoxide Toxicants
18%	Petroleum Distillates (Food Grade)
36%	CFC-11
44%	CFC-12

Minor amounts of methylene chloride and/or iso-butane have also been included in some products as less costly alternatives.

Two other approaches were proposed but rejected.

The petitioning marketers were not interested in providing total release or meterspray products to these food-handling establishments and suggested that EPA exclude those forms from the exemption they were seeking.

Aircraft fumigation is used to prevent the entry of unwanted insects into the United States. In some instances, the products also contain a disinfectant ingredient to reduce surface-contact bacteria, molds, mildew, yeasts, rickettsia, virus, and other micro-flora.

Regulations vary among countries, but many require fumigation with nonflammable formulations in aerosol dispensers, both at their home ports of entry and in aircraft of their register. In the U.S., the American Pilots Association has established a code, accepted by the carriers, that only nonflammable aerosols shall be used for fumigation. This is also a regulation of the Civil Aeronautics Board.

Aircraft are not made explosion-proof, and the possibility exists that excess, localized spraying in a sensitive area by a poorly informed person could lead to a fire, although not to an explosion (9.6 g of hydrocarbon propellant is capable of bringing 55 U.S. Gallons of air to the flammable range).

In the petition for exempting these pesticides from the regulation, the option of a water-based product was considered, but it was rejected for three reasons: 1) it still contained 30 to 32% hydrocarbon propellant, although it was technically nonflammable by standard test methods then in use; 2) the heavier particle size caused fall-out to affect the passengers and polished surfaces; and 3) because of the fall-out, the odor of the toxicants was more persistent than with true space sprays.

The exemption allowed for tobacco barns and warehouses has apparently been liberalized to cover tobacco warehouses and food warehouses if they contain tobacco products. Nevertheless, this is a de minimus application, and some users have converted, because of economic incentives.

The petition claimed that tobacco leaves, dusts, and granules are both costly and very flammable; thus, flammable sprays should not be used in buildings that contain these materials. Tobacco in the dried state is also highly absorbent and would pick up a certain amount of the emulsifiers, inhibitors, and other ingredients of water-based flying insect sprays, adulterating the tobacco.

Metered-Dose Inhalant Drugs (Steroids, ergotamine tartrate and adrenergic
bronchodilator types.)

This category represents about 4% of the U.S. domestic aerosol unit
volume, although the products are very small in size. The usual range is from
about 0.5 Av.oz. (14 g or 10 mL) to about 0.75 Av.oz. (21 g or 15 mL). The
CFC content is 87 to 96 percent. In general, 70 to 80% of the CFC content is
CFC-12. The other CFCs are CFC-11 and CFC-114. A typical formulation
follows:

1.5%	Solid Drug (Generally powdered to 2-10 microns)
1.5%	Excipients
12.0%	CFC-11
10.0%	CFC-114
75.0%	CFC-12

The less common solution types use about 10% anhydrous ethanol as a co-
solvent; about 85% or more CFC-12 is used in such products to achieve the
desired break-up of the spray. Unless the particles are within 0.5 to 10
microns in size, they will not penetrate beyond the extrusile cilial area and
up to the alveolar tubes and sacs where pulmonary abnormalities may exist.
For example, Al-buterol, used in about half of all these products, requires
deep penetration to be effective.

To reduce thermally sensitive powdered drugs into the 2- to 10-micron
particle size distribution, they may be mixed with an appropriate non-solvent
liquid of high volatility into a slurry and then ground or milled. As
frictional heat is generated, it is instantly dissipated into solvent warming
or volatilization. None of the drug houses has explosion-proof equipment in
their FDA-approved processing facilities--nor do they have the flammable gas
detection devices; blow-out walls; electroprotective systems; multi-stage,
high-intensity ventilation; and other systems that would be needed if the
slurry liquid were highly flammable [such as n-pentane (B.P. = 98°F) or isopentane (B.P. = 86°F)]. Consequently, the slurrying agent must be non-

flammable and available at a high level of purity. CFC-11 (B.P. - 74°F) fulfills this requirement.

In some preliminary studies of a CFC-11/hydrocarbon system, the hydrocarbon propellant was found to have a bad (stinging, oily) taste, especially if the product was designed for oral entry, instead of nasal inhalation.

In 1977 and 1978, the pharmaceutical houses advised the FDA (Drug Division) of the above facts, stating that no alternative slurrying agent was available for CFC-11, and that non-CFC propellants were inimical to product organoleptics and performance. They also reported that, even if nonflammable and otherwise acceptable alternatives were available, they would require three to five years of developmental work, followed by a five-year study by the FDA before marketing approval could be obtained. Without a continuing supply of CFCs, these life-saving and remedial drugs would no longer be available.

Contraceptive Vaginal Foams

These drugs were marketed after extensive testing by pharmaceutical firms and by the FDA (Drug Division). Testing of an alternative formula would have to proceed almost from the beginning and would take an estimated seven years to complete. The FDA favored the use of CFC-12/114 propellant blend over the A-46 (iso-butane/propane) blend; the New Drug Amendment (NDA) granted in the late 1960s was based only on the CFC formula option.

Vaginal foams are not a life-saving product. The exemption was granted because, while alternative propellants were available, they could not be used until many years of retesting had been conducted. Also, the market for the products was rather small, and only about 6.4 g of CFCs were used per average 3 Av.oz. dispenser. Had the usage level been higher, the FDA might have not granted the exemption, feeling that other routes to contraception were available.

Hair Restorers

During 1988, however, the FDA granted an NDA to the Upjohn Company for a hair restorer product. The marketer provided information to the agency on products using both CFC and A-46 (iso-butane/propane blend) propellant options. Despite the Montreal Protocol and heightened concerns about the stratospheric ozone layer, the FDA selected only the CFC option for marketing.

The product is a quick-breaking foam, is quite small in size (2 Av.oz.), and contains less than 10% CFC blend. Thus, it is extremely similar to the contraceptive vaginal foam in product/package characteristics. Ultimately, it could capture a substantial market.

Aerosols for Aircraft (Maintenance and Operation)

Little information is available on these aerosols. Along with the Flying Insect Fumigants discussed above, these aerosols include lubricants, cleaners, and other industrial items. Considerations such as safety, high value, and flammability potential—the largest aircraft can hold almost a tank truck-sized load of aviation fuel—have led the American Pilots Association (APA) and related organizations to recommend the restriction of service aerosols to nonflammable options.

Two nonflammable options for a lubricant aerosol are the following:

CFC Type:	2%	SAE 30 Motor Oil - Winterized
	5%	Modified Dimethylsiloxane (Silicone Oil)
	23%	1,1,1-Trichloroethane
	40%	CFC-11
	30%	CFC-12

Alternative:	2%	SAE 30 Motor Oil - Winterized
	5%	Modified Dimethylsiloxane (Silicone Oil)
	83%	1,1,1-Trichloroethane
	5%	Odorless Mineral Spirits (Flash Point 129°F (54°C) Set-A-Flash Closed Cup)
	5%	Carbon Dioxide

Following are two options for spray paints for touch-up and corrosion control:

| CFC Type: | 55% | Aerosol Paint Concentrate (Generally Acrylic) |
| | 45% | CFC-12 |

Hydrocarbon: (flammable)	55%	Aerosol Paint Concentrate (Generally Acrylic)
	15%	Acetone or Methylene Chloride
	30%	Hydrocarbon Blend A-85 52% Propane and 48% Iso-butane

Although APA would not approve of the hydrocarbon version, the Factory Mutual Research & Engineering Company has shown that the two formulations are virtually indistinguishable in terms of intrinsic flammability.

Military Aerosols (For continuing military preparedness)

The U.S. Armed Services purchases about 1,000,000 pounds of CFC-12/CFC-11-propelled insecticides a year by contract. Most are Flying Insect Sprays for troops, but some are specialty items, such as Wasp and Hornet Sprays with CFC-113, designed to protect maintenance personnel working on power stations and utility lines. Significant numbers are exported to military bases overseas.

The Armed Forces also purchases various other aerosol products, and may at their option select CFC formulations, apparently without specific justification. The number of CFC aerosols is thought to be relatively low, and their use is restricted to military aircraft maintenance, computer de-dusting, metal flaw finding, magnetic tape sprays, and specialty lubrication.

Diamond Grit Sprays

Very little information could be found, except that the market is minuscule.

DuPont, Allied-Signal, and Racon had no record of selling CFCs for such an application. No one could imagine using such a product except for diamond "sand blasting" extremely hard metallic surfaces to micro-etch them.

CFC-115 For Puffed Food Product Aeration

DuPont, which is the only remaining domestic supplier of CFC-115 for this use, has not sold the product to a whipped cream manufacturer or food company in many years.

The original whipped creams consisted of water dispersions of either natural or synthetic whipped cream ingredients, pressurized with CO_2 and N_2O (or the White's Mixture of these food grade propellants). For economic reasons, formulations were then limited to synthetic whipped creams with nitrous oxide (N_2O) propellant. The limited solubility of the N_2O gas forced marketers to fill cans to only about 55 to 60% by volume and to use maximum pressures of N_2O. Otherwise, the product became "soupy" near the end of the can.

From 1968 through 1971, when food aerosol dispersions increased dramatically in the aerosol industry, DuPont developed their Freon (Food Propellant) C-318 (chemically, perfluorocyclobutane), which could be added to the N_2O at the rate of 3 to 6 g per 9 Av.oz. can to give a much more uniform whip throughout the life of the dispenser. It also allowed 2 to 3 additional

Av.oz. to be included in the can, at the penalty of slight "soupiness" near the end.

The problem was that the C-318 cost $7.50/lb. A typical use of 4.5 g per can increased the factory cost by $0.20 per can, and even more if extra product were added to the can. Since the new cans looked the same size as the original ones, consumers bought the lower-priced original formulas, and C-318 production ended.

DuPont then obtained FDA approval for CFC-115, which was well known but never used for aerosols because it cost $1.50/lb. CFC-115 was tried in several products during the 1977 to 1978 petition period. Marketers based their petitions on the argument that their sales would seriously decrease without the "bail-out" attributes of CFC-115, that the propellant had only one chlorine atom and should pose a reduced threat to the ozone layer, and that only about 5 grams were needed per can. At the time, roughly 38,000,000 units of whipped creams were sold a year, or 190 metric tonnes per year of CFC-115, if every marketer used the propellant.

The marketer team also advised the FDA that no other propellants were practical. Freon C-318 was unavailable except in pilot quantities, the Food Grade hydrocarbons gave the product an oily, slightly biting off-taste, and nitrogen was too insoluble.

After the petition was granted, with limits on product type and usage levels, the price of CFC-115 increased moderately. Marketers began to look at another option. The American Can Company had recently introduced their Rim Vent Release (RVR) mechanism on the top double seam of three-piece aerosol cans, which opened up a number of apertures if the can became dangerously overpressurized. The Department of Transportation (DOT) normally limits aerosol pressures to 180 psig at 130°F if the cans were ordered with RVR fixtures. The product was nonflammable and not otherwise hazardous if it did escape the container. Ultimately, DOT granted an industry petition, which allowed marketers to insert 17% more N_2O gas into the product. This reduced the "soupiness" problem and slowly led to the complete elimination of CFC-115 in these products.

At least two firms are testing CFC-115 in anhydrous "edible whip" products of the general type recently patented by the Ring Chemical Company in New Jersey. It gives them a better taste than the 3% Propane mentioned in the patent. One firm is considering use of CFC-115 until HFC-134a (CH_2F-CF_3) comes onto the market and can perhaps be approved as a Food Grade propellant. Propane may also be used, as a last resort.

RATIONALE FOR EXCLUDED USES OF CFC AEROSOLS

Drain Openers

The aerosol drain opener was developed by Glamorene, Inc. (Secaucus, NJ) about 1972. The actuator consisted of a 2 and 11/16 inch plastic hemisphere with the exit hole at the apex. The unit was upended over the drain hole of a clogged sink and pressed downward, opening a valve to allow CFC-12 gas into the drain pipe, which normally acted to blow the clog out of the gooseneck. Problems were encountered when drain liquids were blown out of the other drain in double sink installations. Many bathroom sinks had side orifices to prevent overflow, and these sometimes acted as pressure outlets as well. Pipe connectors sometimes separated.

The product consisted of 100% CFC-12. It did not meet criteria such as life-safety, high value in use, no available alternative, and minimum consumption of CFC gas-liquids.

Eventually the high cost, excessively large can size, infrequency of use, and the many problems caused consumers to discontinue purchasing it. The franchise was sold to another marketer, and sales eventually ceased.

Microscope Slide Cleaners

A blend of about 75% ultrapure CFC-113 and 25% CFC-12 is used to flush across the microscope lens while it is still in the microscope holder. Loose dry specimens, specimens under oil, dusts, and other objects are conveniently flushed off, after which any remaining CFC-113 quickly evaporates. This is a great time-saver for many institutional facilities where repetitive assays or examinations are required.

Chewing Gum Removers

Flattened lumps of chewing gum are difficult to remove from carpeting and other floor coverings, but the task is greatly facilitated by using a chewing gum freezant spray of CFC-12, which acts to draw down the gum temperature to about -60°F (-51°C) or lower by successive evaporations off the surface. The "pancake" then becomes very brittle, able to be fractured or broken up with a sharp blow. The pieces can then be removed while still frozen.

This aerosol product was launched after the 1978 restrictions or bans. It is used mostly in institutions and accounts for about 350,000 lbs. of CFC-12 per year. As of July 1989, at least two major CFC suppliers refused to sell any more CFC-12 for this application.

Boat Horns

The dispensers and associated horn devices are made by specialty firms. Arguments for continuing the use of 100% CFC-12 in such packages are 1) the sense of weightiness (15 Av.oz. per can) that connotes a good product; 2) the nonflammability of the product (sometimes large amounts are released at one time); and 3) marketers usually manufacture their own products in facilities that are not explosion proof and where they could not safely produce the less costly hydrocarbon propellant alternative products.

Most of these marketer/fillers produce a considerable variety of aerosol products, all of them pressurized with CFC propellants or containing 100% CFCs. By September 1989, several manufacturers had switched to using HCFC-22 in heavy walled (nonaerosol) cylinders.

Halon-type Fire Extinguishers

The portable dispenser area of this market is divided into two segments: the aerosol type and the cylinder type. The aerosol is limited to two sizes: the twin-pack of two 4 Av.oz. cans in a belt-carrier canister, with a common actuator, and the 12 to 15 Av.oz. pack, in a red enameled, Mylar-labeled aluminum container, coded and approved for use by Factory Mutual R&E Laboratories, Underwriters Laboratories, or one of the three other product testing laboratories recognized as reliable. The aerosols nearly always use a blend of 20 to 25% Halon 1301 ($CBrF_3$) and 80 to 75% Halon 1211 ($CBrClF_2$). The objective is to obtain a pressure that will allow the aerosol to pass specific fire tests, yet that will not exceed the "DOT Specification 2Q" specification of 180 psig (maximum) at 130°F (54°C), and that is considered "self propelling" (not requiring a pressure gauge).

The cylinder or tank portion of this market is several times larger. Several firms sell six or more product sizes, offering the aerosol size for car or boat and the larger sizes for home use.

The Halon is considered "the product" since it provides the fire extinguishing action. It is considered a life-safety product. Unfortunately, bromine is also 3 to 10 times more able to destroy ozone than chlorine (Cl), depending on the Halon molecule it is in.

Intruder Alarm Devices (For cars, trucks, and homes)

These products are high-intensity (100 to 110 decibel) horns triggered electrically by various sensing devices. Large amounts of gas are discharged in each alarm cycle to achieve a persistent sonic effect. Because of this, nonflammable gas is preferred, especially for indoor locations. The cost premium for the CFCs (over the hydrocarbon alternative) represents a small increment of the overall device.

Skin Chillers (For medical purposes)

These aerosols are typically packaged in aluminum cans, filled to 8 to 10 Av.oz. with CFC-12 or a blend of CFC-12 and CFC-114. They are used by medical personnel to perform minor topical operations, such as the removal of a splinter, a wart, a non-melanomic skin cancer, and so forth, by deadening the afflicted area. It avoids the use of injected Novocaine and similar anesthetics. Chillers have also been used to shrink fingers and facilitate the removal of rings that have become too tight.

Using flammable gas alternatives in a hospital or clinical setting would not be approved by the medical profession. Products known to have such propellants would not be used, despite their lower cost. Dimethylether is the most effective propellant for this application, but it is somewhat flammable.

Polyurethane Blowing Agent

A typical formula for this type of aerosol product is as follows:

45%	Toluene di-isocyanate
20%	Polyethylene Glycol (PEG) Derivatives
32%	CFC-12
3%	Dimethylether (DME)

The dimethylether is present only to bind up residual moisture that can otherwise act to catalyze the reaction between the pre-polymer and the PEG derivatives. The fact that it is a propellant is incidental. The CFC-12 is used to ensure production of an essentially nonflammable, stable foam. In cursory studies with foams containing hydrocarbon propellant blowing agents, ignition was instant and was followed by very rapid burning--with the production of modest amounts of cyanides, cyanogen, and related toxic substances.

Tire Inflators

Approximately 75% of the usual tire inflator consists of a water dispersion of resin and ethylene glycol, and a trace of amphoteric surfactant; 25% is a hydrocarbon blend A-46 (iso-butane and propane).

When a tire is reinflated with these products, the gas space will then contain about 35 to 65% by volume of flammable hydrocarbon gas. This is well above the Upper Explosive Limit (UEL) of about 8.6 v% for propane/butanes and air, but in subsequent activities (some directed on the label) this volume may be diluted with air into the flammable range of 1.9 to 8.6 v%. Tires in this state have been subjected to removal, to adjacent rim welding repairs, and to other incidents that resulted in internal combustion, fragmentation of the tire, and severe, often fatal, injuries to the operator.

In 1986 one such marketer met with representatives of EPA and the Consumer Products Safety Commission (CPSC) in Washington, DC, proposing that the agencies permit the substitution of the hydrocarbon propellant by 55% CFC-12/114 (40:60). The agencies confirmed that they had the right to do this because the CFCs would constitute a part of the product and because they had inflating action. Within two months the revised product was on the market.

Lawsuits increased, and during late 1988 another marketer developed a CFC-based tire inflator formula, but before going to market, the marketers estimated that for every million pounds they purchased 18 skin cancers and 0.3 deaths would occur annually. The firm did not market the product.

The CFC product is technically inferior and more costly than the standard hydrocarbon types; therefore, it has only been marketed by one firm.

Foams, Whips, and Puffs

From 1981 to 1986 a few small specialty firms produced CFC products of these types. For example, a firm formulated and packed an ablative, very heavy foam using CFC-12/114 (40:60). Finger-rings containing gemstones could be partly embedded in this foam while the back-side was being welded during re-sizing operations by jewelers. The stones remained cool and did not become strained or cracked. The self-filler had no explosion-proof facilities (in his basement) and was concerned about the flammability of foams that carried hydrocarbon propellants.

No other evidence was found of CFCs currently being used in the U.S. in non-drug aerosol products of this type.

Medical Solvents (Bandage adhesive and adhesive remover)

Two firms purchase a concentrate from the Dow Corning Corporation that contains a silicone-based adhesive dispersed in CFC-113. Contract fillers pour the concentrate into cans, seal them, and pressurize the contents with carbon dioxide (CO_2).

Dow-Corning scientists say that alternatives to CFC-113 are uniformly unacceptable for the product. For example, the 7 to 8% of free-radical reaction inhibitors in 1,1,1-trichloroethane have unknown effects when brought into contact with open wounds.

The adhesive remover is required since the breathable silicone film may bond the innermost layer of cotton gauze bandage to the skin surface, preventing removal. Acetone, ethanol, and many other solvents are either ineffective or present toxicological problems. Thus, a self-pressurized adhesive remover is required. This consists of about 95.4% CFC-113 and 4.6% CO_2, in a fill weight of 6.0 Av.oz. per can.

SUMMARY

Approximately 28 individual product types and groups have been or are being produced in aerosol formulas that contain CFCs. This does not include CFC refrigeration/air conditioning refill products or ethylene oxide/CFC-12 gas sterilants, since these are not considered in aerosol exemptions or exclusions.

Background data on each product or product group is provided, including how the CFC component functions. Industry's interest in preserving the CFC ingredient(s) is explained. In the excluded categories, petitions were not required, but meetings were often held, nonetheless, to firmly establish that companies could go to market with CFC-based products falling into these categories. The primary reasons for requesting exempted status were the unavailability of substitutes, the long time delays while obtaining an amended NDA from the FDA, solvency and purity profiles (especially for CFC-113 uses), life-saving potential of the product, and regulations in hospitals, aircraft organizations, etc. against the use of flammable propellants in aerosols.

During the 1977 to 1978 transition period, no nonflammable liquid propellant alternatives to CFCs were toxicologically approved and commercially available for use. Today the situation has changed, with the clearance of HCFC-22 and certain blends of HCFC-22/142b, and the forthcoming availability of HFC-134a, HCFC-123, HCFC-141b, and HCFC-124.

As may be anticipated, some exempted or excluded aerosol products are no longer in use or have been replaced with ones that contain alternative propellants. However, inhalant and solvent type products are steadily growing in sales volume. See Section 3 for current U.S. consumption of CFCs and Section 4 for current and future alternative formulations for CFC aerosols. Section 5 presents a discussion of the costs of making these substitutions to alternative formulations.

3. Current U.S. Consumption of CFCs

The main sources of information for this section are CFC manufacturers. Product formulas, average product net weights, and the differences in the sales volumes of similar products are estimated. The production volume of the aerosol industry in units per year for 1988 is determined from valve sales and from comparisons with the survey data published in 1987 by the Chemical Specialties Manufacturer's Association (CSMA) and the Can Maker's Institute (CMI). Categorical volumes are then derived by extending and augmenting the detailed CSMA data with specific data from industry contacts.

DISCUSSION OF THE DATA

Table 3 shows published data (5) illustrating an analysis of the major categories of the domestic CFC market. Table 4 shows a number of product categories of aerosols in which CFCs are still being used, or have been used since 1978. The U.S. CFC aerosol market in 1986 was 24,000,000 pounds. It grew by 3.9% in 1987 to 25,000,000 pounds. The CFC aerosol market in 1988 was projected to be about 25,500,000 pounds (as of October 1988).

The following paragraphs explain the information presented in Table 4.

Source of the Data

The source of the data obtained on CFC quantities and units filled is the aerosol industry, which defines aerosols in the same way as the U.S. Department of Transportation: pressure-resistant containers with a capacity of 50 cubic inches--819.35 mL or 27.71 fluid ounces. Larger containers, designated as cylinders or tanks, do not have the aerosol exemption, and (in general) are filled and marketed by firms other than those in the aerosol

27

TABLE 3.   THE FULLY HALOGENATED U.S. CFC MARKET OF 1986
(CFC-11, -12, -113, -114, AND -115) (5)

PRODUCT TYPE	PERCENT	MM POUNDS
Refrigerants	40	290
Blowing Agents	28	200
Cleaning Agents	20	145
OTHERS	12	85
Liquid Freon Freezant Aerosols Etchants Sterilants		
	100	720

NOTES (6):

a.   Liquid Freon Freezant (LFF) is 100% CFC-12.

b.   Etchants are perfluoro- or highly fluorinated CFCs, mainly CFC-114,
but including CFC-113, CFC-115, and FC-116. They are deliberately
degraded by plasma arcs to produce HF for the etching of electronic
parts.

c.   The sterilants consist of the U.S. Patented, marginally nonflammable blend of 12% ethylene oxide (EO) and 88% CFC-12. Most is
sold to medical facilities in large cylinders, but some is marketed
in 12 Av.oz. aerosol containers.

d.   The "Cleaning Agents" are 100% CFC-113.

e.   mm - million

TABLE 4. AEROSOL MARKET PENETRATION OF CFC-CONTAINING PRODUCTS - 1988

PRODUCT DESCRIPTION	CFC CONSUMPTION (MM lbs)	AVERAGE CONTAINER SIZE (Fill wt - lbs)	PRODUCT VOLUME OF CFC VERSIONS (MM Units)	PRODUCT VOLUME FOR ALL VERSIONS (MM Units)	PERCENT OF PRODUCT CATEGORY USING CFCS
EXEMPTED PRODUCTS[a]					
Mold Release Agents	1.5	1.11	1.32	2.79	47.3
Lubricants for Electric/Electronic Equipment	1.9	0.76	2.32	12.8	18.1
Lubricants for Pharmaceutical Pill-making	1.0	0.86	1.93	1.93	100.0
Solvent-cleaners and De-dusters for Electric/Electronic Equip.	6.0	0.51	10.9	19.8	55.1
Mercaptan Warnings	<0.1	0.95	<0.09	<0.09	100.0
Intruder Alarm Device	0.6	0.47	1.21	1.21	100.0
Flying Insect Sprays for Commercial Food Handling Areas	<0.1	0.76	<0.12	3.20	<3.7
Flying Insect Sprays for Aircraft	0.3	0.48	0.58	0.58	100.0

(Continued)

TABLE 4. (Continued)

PRODUCT DESCRIPTION	CFC CONSUMPTION (MM lbs)	AVERAGE CONTAINER SIZE (Fill wt - lbs)	PRODUCT VOLUME OF CFC VERSIONS (MM Units)	PRODUCT VOLUME FOR ALL VERSIONS (MM Units)	PERCENT OF PRODUCT CATEGORY USING CFCS
Flying Insect Sprays for Tobacco Barns	<0.05	0.76	<0.06	0.06	100.0
Metered Dose Inhalant Drugs: --Steroid Drugs --Ergotamine T. Drugs --Adr. Bronchodilator	3.9	0.033	107.0	107.0	100.0
Contraceptive Vaginal Foams for Human Uses	<0.1	0.16	<6.94	<6.94	100.0
Aircraft Maintenance and Operation Sprays	0.5	0.76	0.61	12.5	4.9
Military Aerosols	1.4	0.68	1.96	35.0	5.6
Diamond Grit Sprays	<0.05	0.76	<0.06	<0.06	100.0
CFC-115 for Aeration for Puffed Food Prods.	0	0	0	0	0

(Continued)

TABLE 4.   (Continued)

	CFC CONSUMPTION	AVERAGE CONTAINER SIZE	PRODUCT VOLUME OF CFC VERSIONS	PRODUCT VOLUME FOR ALL VERSIONS	PERCENT OF PRODUCT CATEGORY
EXCLUDED PRODUCTS[b]					
Tire Inflators	1.0	0.93	1.65	17.02	9.7
Polyurethane Foam	3.1	1.01	9.17	9.17	100.0
Chewing Gum Removers	0.3	0.59	0.47	0.47	100.0
Drain Openers	<0.05	0.70	<0.07	<0.07	100.0
Chillers (Medical)	0.3	0.39	0.71	0.71	100.0
Boat Horns	1.9	0.77	2.28	2.28	100.0
Non-E/E De-dusters	0.4	0.43	0.86	0.86	100.0
Microscope Slide Cleaners & Related Products	<0.05	0.39	0.12	0.12	100.0
Solvents (Medical)	0.6	0.51	1.09	1.09	100.0
Contingency Products, Including Unauthorized Uses	0.2	0.50	0.41	--	--
	25.4				

(Continued)

TABLE 4.   (Continued)

PRODUCT DESCRIPTION	CFC CONSUMPTION (MM lbs)	AVERAGE CONTAINER SIZE (Fill wt - lbs)	PRODUCT VOLUME OF CFC VERSIONS (MM Units)	PRODUCT VOLUME FOR ALL VERSIONS (MM Units)	PERCENT OF PRODUCT CATEGORY USING CFCS
OTHER PRODUCTS					
Aerosol Size Halon Fire Extinguishers 1301/1211 (Includes 30% Imported)	3.8	0.83	4.24	4.24	100.0
Refrigerator/Air-conditioner Refills (12 and 114)	67.7	0.96	70.5	70.5	100.0

NOTE[a]:   These products are in the exempted category. The volume totals 17.5 MM lbs.

NOTE[b]:   These products are the excluded and non-regulated types, in which CFCs form part or all of the concentrate; hence, they are not considered aerosol propellants. The volume, not including the "other" products (Halons and Ref/AC refills), is 8.0 MM lbs.

industry. The aerosol surveys by the Chemical Specialties Manufacturer's
Association, Inc. (CSMA), the Can Maker's Institute (CMI), Precision Valve
Corporation (PVC), et al. consider only those aerosols of capacities equal to
or less than 50 cubic inches. Butane lighter fluids are not considered to be
aerosols; however, CFC refill units within the size limitation are.

Although some exemptions have been granted for products marketed both as
cylinders and as aerosols, as defined above, for the purposes of this report,
the cut-off is the 50-cubic inch level; larger sizes are not considered in any
detail.

Development of Numbers by Exact Product Type

When developing numbers for the column titled "Product Volume for All
Versions (MM Units)," only units of the exact product type were considered.
For example, the "Mold Release Agents" line refers to release agents for molds
used in the production of plastic and elastomeric materials. Figures for mold
release agents used for toy metal (tin-bismuth) soldiers, cookie forming, ice-
cube-makers, some candlemaking, and other applications are not included, even
though a small quantity of CFC is used in these areas.

Similarly, "Flying Insect Sprays for Aircraft Fumigation" were not
compared with the 106,400,000-unit overall flying insect aerosol market.

Flying Insect Sprays

The use of flying insect sprays based on CFC formulations in commercial
food preparation areas has dwindled to almost nothing. Previous formulas have
been replaced by water-based, hydrocarbon-propelled formulas, supplemented by
anhydrous, hydrocarbon-propelled types using time-metering canisters attached
to walls or posts.

A major concern was restricting these sprays to food handling areas.
Employees were apt to spray bathrooms, dressing areas, and other unauthorized
locations. The higher cost was a further concern.

Although meter-sprayed aerosols are not exempted, the new formulas have been slow to replace the meter-spray CFC-12 types because the meter valve supplier could not readily produce a complex valve modification that would spray 40-45 mg shots of the low-density hydrocarbon formulas in place of the usual 100-110 mg bursts of the standard CFC compositions (7).

Aerosols for Military Use

The "Military Aerosols" category includes two sub-groups: the flying insect spray (CFC-type) and electric/electronic maintenance and testing products (also CFC-type). The U.S. government orders 1,500,000 10-Av.oz. units formulated with 2% Toxicants and 98% CFC-12/11 once a year. Depending on insect populations during any year, from about 1,250,000 to the full 1,500,000 units are released for shipment from the sole supplier.

These formulas account for up to 1,020,000 lbs of CFCs. In Table 4, the remaining 380,000 lbs of CFCs are for electric/electronic maintenance activities around sonar, radar, radio, servomotor, computer, airplane controls, and other sensitive, costly, or sparking types of equipment.

Metered Dose Inhalant Drugs

The three classes of meter-spray inhalant aerosols used for medical purposes can be defined as follows:

- Metered-dose steroid human drugs for oral inhalation;

- Metered-dose steroid human drugs for nasal inhalation; and

- Metered-dose adrenergic bronchodilator human drugs for oral inhalation.

Information from the Haskell Laboratories Division of DuPont shows the equivalence of nasally and orally inhaled drugs of the steroid types; they are generally identical except for the plastic nose or mouthpiece.

Table 4 of the present report only gives a total figure for all metered-dose inhalant drugs. The breakdown of these drugs is as follows:

Steroid drugs--oral	20 (% units)
Steroid drugs--nasal	10 "
Ergotamine Tartrate vasoconstrictor drugs	3 "
Bronchodilator drugs--oral	67 "

The Ergotamine Tartrate types represent less than one percent of CFCs used in this category. This inhalant is a cranial vasoconstrictor, used for migraine and prodrome control. Two new metered-dose ethical drug products. a hair restorer (quick-breaking foam) and a curative pulmonary spray for bronchial pneumonia--so far only FDA-approved for AIDS patients--are reported to contain CFC propellants. The volumes for these products are unavailable. since each is made by only one firm and the market results would be too revealing. The hair restorer was released in 1988 and the other was conditionally approved in early 1989.

Additional details on metered-dose CFC-containing drug aerosol products appear in Appendix A.

Conversion of Pounds of CFCs Consumed to Aerosol Units Sold

When calculating aerosol units offered for sale from pounds of CFCs consumed (and vice-versa) it is seen that:

• Aerosols are regulated as "delivering systems" for net weight purposes, and must therefore be filled to an average of 0.10 to 0.25 Av.oz. over labeled weight, according to size.

• An average of about 8% total propellant is lost during the handling and filling process. This includes leakages, machine discharges. and the 1.5 to 2.5% of all aerosol units lost through process leakage, bad quality, testing, samples, pilferage, etc.

- Not all CFC aerosols are 100% CFC. Examples of CFC percentage levels in other products are:

Flying Insect Killers (Military)	98%
Lubricants and Mold Releases	95%
Inhalants	94%
Contraceptive Foams	8%

Conversion equations then become:

$$\frac{\text{CFC Usage (lbs)}}{\text{Actual fill (lbs) x CFC Loss Factor x CFC Fraction}} = \text{Units}$$

U.S. MARKETERS AND FILLERS OF CFC PRODUCTS

A substantial number of marketers and contract fillers currently produce the estimated 5% of all aerosols that still contain CFC-type propellants. Some of these firms are listed in Table 5.

TABLE 5.   MARKETERS AND FILLERS CURRENTLY PRODUCING
CFC-TYPE AEROSOLS IN THE U.S.[a]

Tradenames	Company Name	City Address
Fluorosolv	Kem Manufacturing Corp.	Tucker, GA
Flux-Off	Chemtronics, Inc.	Hauppauge, NY
En-Rust, Cobra, Aeroduster	Miller-Stephansen Chemical Company	Danbury, CN
Aervoe	Aervoe-Pacific Company	San Leandro, CA
All Four, Fault Finder	Crown Industrial Products Company	Hebron, IL
AudioTex & Calectro	G.C. Electronics Division	Rockford, IL
Chargette	Airosol Company	Neodessa, KS
Cold Spray	Zee Medical Products Co.	Irvine, CA
Hollister	Hollister Medical Products	Chicago, IL
JC	Larson Laboratories, Inc.	Erie, PA
Contax & Grime-Solv	Stewart-Hall Chemical Corp.	Mt. Vernon, NY
Delfen Foam (Vaginal)	Johnson and Johnson, Inc.	Raritan, NJ
Exit (Chewing Gum Rem.)	Avmor, Limited	Montreal (Canada)
Drain Power (Obsolete?)	Airwick Industries, Inc.	Carlstadt, NJ
ReleasaGen (four items)	ReleasaGen Manufacturing Co.	Delano, MN
Minus 62 (Freezant)	Tech Spray, Inc.	Amarillo, TX
Lectro-Safe	DuBois Chemicals, Inc.	Cincinnati, OH
Koromex Foam (Vaginal)	Holland-Rantos Co., Inc.	Trenton, NJ
Lubri-Bond	Electrofilm, Inc.	Vanencia, CA
Mace (Several types)	Smith & Wesson Co.	Springfield, MA
Mace (Other types)	General Ordinance Equip. Co.	Springfield, MA
Mold-Ease & Solv-Clean	Chem-Pak, Inc.	Winchester, WV
Quick Freeze	Sentry Chemical Company	Stone Mountain, GA
Freez-It, Lubrite, etc.	Workman Electronic Products	Sarasota, FL
Industrial Freezant	Rolmar: The Supply Corp.	Lake Geneva, WI
Freez'n Check & 70 PSI	ChemTronics, Inc.	Hauppauge, NY
Steel-One	Madison Bionics, Inc.	Franklin Park, IL
Seal Aid & Spra-Dri	Orb Industries, Inc.	Upland, PA
Zapper (Personal Protection)	Safety & Security Company	Harleysville, PA
(Contract Filler)	Aerosol Systems, Inc.	Macedonia, OH
(Contract Filler)	Chase Products, Inc.	Maywood, IL
Dri-Slide	Dri-Slide, Inc.	Freemont, MI
(Contract Filler)	Armstrong Laboratories, Inc.	West Roxbury, MA
Electro-Freez, Falcon, etc.	Falcon Engineering Co.	Mountainside, NJ
Flaw Finder	American Gas & Chemical Co.	Northvale, NJ
Benvenue	Benvenue Laboratories, Inc.	Bedford, OH
Globe Spray (For endoscopic work)	Globe Medical Instruments, Inc.	Clearwater, FL
Makiki, Voltex & Term-Out	Makiki Electronics Company	Hauula, Hawaii

[a]Inhalant drug producers are not listed.

4. Suggested Alternative Formulations for Exempted and Excluded CFC Aerosols

INTRODUCTION

The CFC propellants in exempted and excluded aerosols are currently preferred for a variety of reasons. Such reasons include their nonflammability, high purity, unique solvent characteristics (CFC-113 in particular), or the fact that they have been made an essential part of New Drug Amendments (NDAs) for pharmaceutical meter-spray aerosol products.

Other propellants or blends may exhibit some of these properties. Since 1983, a set of four alternatives to CFCs has been commercialized, and another set of four is expected to come onto the market in 1992 or 1993 if the results of toxicological testing programs continue to be favorable. The physical properties of the eight alternatives are summarized in Table 6. All the presently available alternative propellants are gases at room temperature. Boiling points are 14°F or lower; therefore, none can be considered a direct replacement for CFC-11 (B.P. = 73.3°F) or CFC-113 (B.P. = about 120°F), which are liquids at room temperatures.

Although HCFC-22 is nonflammable and is commercially available, teratogenic uncertainty has discouraged marketers of metered-dose pharmaceutical inhalant aerosol sprays. The extreme solvent activity of dimethyl ether is an additional concern for those firms marketing drug products in solid suspension forms because of Ostwald Ripening effects that change particle size distributions.

In this section, two or more formulation options will be presented for a number of products that currently use regulated CFCs. Some have a relatively greater potential for stratospheric ozone depletion than others. Although the

TABLE 6. CURRENT AND POSSIBLE FUTURE ALTERNATIVE AEROSOL PROPELLANTS

PRODUCT:	Currently Available				Available in 1992 or 1993			
	HCFC-22	Dimethyl-Ether	HFC-152a	HCFC-142b	HFC-134a	HCFC-141b	HCFC-123	HCFC-124
Formula	$CHClF_2$	$(CH_3)_2O$	CHF_2-CH_3	$CClF_2-CH_3$	CF_3-CH_2F	CH_3-CCl_2F	CF_3-CHCl_2	$CF_3-CHClF$
VP (psig) @ 70°F	121	63	63	29	81	<0	<0	46
VP (psig) @ 130°F	297	174	177	97	202e	17e	25e	125e
Boiling Point (°F)	-41	-13	-13	-14	-16	89.6	83.7	12
Flammability Range, % (LEL-UEL)	NF	3.3 - 18	4 - 17	7 - 15	NF	6 - 15	NF	NF
Solubility in H_2O, wt%	3	35	1.7	0.5	NA	NA	NA	NA

e=estimated

NA = Not available, but low

NF = Not flammable

aerosol industry has received sample amounts of most "future possible alternative propellants," product properties must be estimated when proposing the use of these materials in any future aerosol formulations.

DISCUSSION OF ALTERNATIVE CFC FORMULATIONS

This section will present alternatives to CFC aerosol formulations by 1) discussing the currently exempted and excluded products individually; 2) offering general comments; 3) suggesting alternative formulas; and 4) drawing conclusions.

Factors that were considered when developing the alternative formulations include the following:

- Production of sprays with desired particle size distribution;

- Control of flammability;

- Precautionary use of questionable solvents, such as methylene chloride;

- Minimum changes in anticipated use patterns;

- Maintenance of dispenser and organoleptic stabilities;

- Cost of alternative formulations;

- Availability of the alternative, including Toxic Substance Control Act (TSCA) considerations;

- Pressure limitations;

- Product utility or efficacy for intended uses;

- Toxicological factors; and

- Spray rate and use-up rate optimization.

Mold Release Agents

These products are currently formulated from an approximately 3 to 5% base of lecithin, functional silicone, or other material dissolved in a CFC-11/CFC-12 blend. Product purity, propellant volatility, and nonflammability are key properties. The plastics and elastomers being molded should not be contaminated with such unknown entities as solvents or inhibitors. The base should be delivered to the mold surface with as little loss (bounce) as possible. CFC-11 currently provides the desired high-transfer efficiency.

Two representative CFC-based formulas are as follows:

I.	3%	Lecithin (Soy Bean Source)
	57%	CFC-11
	40%	CFC-12

II.	5%	Silicone (Dimethacone -- 1000 centistokes)
	65%	CFC-11
	25%	CFC-12
	5%	Propane A-108

Both are intrinsically nonflammable.

When traces of inhibitors (such as nitromethane or 1,4-dioxane) are not a problem, an optional formulation such as the following may be suggested:

III.	3%	Lecithin (Soy Bean Source)
	27%	Methylene Chloride - Inhibited
	5%	Ethanol - Anhydrous (Liquid denaturants only)
	39%	HCFC-142b
	26%	HCFC-22

The inclusion of methylene chloride is based on studies by Dow Chemical Company, et.al. showing that the single positive mutagenicity study was (animal) species specific and that the compound does not affect humans. The total

formulation would have a pressure of about 48 psig at 70°F (air-free basis).
It would have a density of about 1.19 g/mL at 70°F, meaning that only about
80% of the present weight of these products could be placed in the filled can.

If methylene chloride is considered unacceptable from the standpoint of
toxicology or solvency, it could be replaced by 1,1,1-trichloroethane -
inhibited (4), as in the following formulation:

IV.   3%   Lecithin (Soy Bean Source)
                37%   1,1,1-Trichloroethane - Inhibited
                36%   HCFC-142b
                24%   HCFC-22

This would bring approximately 3% of decomposition inhibitors (as free-radical
chain breakers, such as nitromethane, etc.) into the formula, since they are
necessary to counter the self-destruction of the 1,1,1-trichloroethane.

The boiling point of 1,1,1-trichloroethane (about 152°F) is high enough
that small amounts could remain for a short time on the die surfaces, depending on conditions, and cause voids in the molded part.

The most costly formulation must await the availability of HCFC-123 or
HCFC-141b. The formulation could be as follows:

V.   3%   Lecithin (Soy Bean Source)
                65%   HCFC-123 (B.P. - 82°F)
                32%   HCFC-22

If a preference develops for HCFC-141b (flammable) over HCFC-123 because
of price or toxicology, a final formula option can be considered:

VI.   3%   Lecithin (Soy Bean Source)
                14%   HCFC-123 (B.P. - 82°F)
                50%   HCFC-141b (B.P. - 90°F)
                33%   HCFC-22

Formulas V and VI would have pressures and liquid densities comparable to those of the present products. Solvency, for the bases, would be excellent. Formula VI could be checked for flammability, as should the possibility of replacing the 14% HCFC-123 with 14% additional HCFC-141b, should this be desired.

Lubricants for Electric and Electronic Equipment

Aerosol lubricants may assume numerous forms, according to the end use. For example, a lubricant for screen doors, sliding aluminum doors, door locks and hinges could consist of the following:

VII.	80%	Specialty Lubricant Blend
	15%	1,1,1-Trichloroethane - Inhibited (4)
	5%	Carbon Dioxide

This lubricant could also be dispensed through a capillary extension tube up to 8 inches long fitted into the actuator orifice.

Electrical applications, as for motors, transformers, switches, and relays, generally require a nonflammable product and a spray delivery. This is especially true of aerosols, where large-scale, spark-producing equipment, such as the commutator ring of a 250 HP motor or dynamo, must be lubricated. CFCs have been used in these lubricants to ensure that the combustibility of the specialty oil itself is suppressed. Some of these products are also used in rooms or areas where smoking and the use of flammable or combustible solvents are prohibited. The composition of one such product follows:

VIII.	5 - 15%	Specialty Lubricant Blend
	65 - 50%	CFC-11
	30 - 35%	CFC-12

The use of hydrocarbons or other flammable propellants is contraindicated for electrical equipment because sparks of sufficient energy (over 0.2 kilojoules) can ignite flammable-range mixtures of these gases and air, harming both operator and equipment. In some instances, spark streams or electrical resistance may act to strongly heat a surface, causing combustible solvents to heat up beyond their flash point, and flammable solvents to heat up beyond their auto-ignition point.

Finally, easily decomposed halogenated solvents may produce hydrogen chloride, phosgene, and other active gases that can act corrosively on hot contact points, resistors, etc. The CFCs (especially CFC-12 and CFC-114) are highly resistant to pyrolysis.

In the electronics area, sprays must be nonflammable because of equipment or environment restrictions; pyrolytic corrosion is also a concern. One of several products is formulated as follows:

IX.   5%   Specialty Lubricant Blend
            65%   CFC-113
            30%   CFC-12

The CFC-113 carries the lubricant to the surface to be treated, allowing a minimum amount to contaminate the air. At the same time it exerts a degreasing, flushing action, removing unwanted vapor fluxer residues and dusts.

Several alternatives to CFC products are available. First, when motors, computer boards, and similar items are not electrically connected, and unless the added solvency of 1,1,1-trichloroethane (4) is a problem, Formula VII can be used in conjunction with a meter-spray valve that will reduce the amount of lubricant sprayed out. For example, a 6 Av.oz. can be made to give about 1700 "shots" of about 100 mg each.

For greater uniformity of the spray pattern during package life, another formula could be considered:

X.   60%   Specialty Lubricant Blend
            12%   1,1,1-Trichloroethane - Inhibited (4)
            28%   HCFC-22

If the presence of 1,1,1-trichloroethane creates a problem, the following more costly meter-spray option should be useful in a "DOT Specification 2Q" aerosol can:

XI. 68% Specialty Lubricant Blend
     32% HCFC-22

Finally, if a product comparable to Formula IX is needed, a counterpart to CFC-113 will present the greatest challenge. The best candidate is the following:

XII. 5%   Specialty Lubricant Blend
      72% HCFC-141B
      23% HCFC-22

The structural similarities of HCFC-141b (CH_3-CCl_2F) and CFC-113 ($CClF_2$-CCl_2F) suggest they may possess a similar solvency potential. Formula XII is of marginally acceptable flammability. This can be corrected, if necessary, by using additional HCFC-22, or by using a meter-spray valve. In time, the replacement of HCFC-22 with HFC-134a could result in a lower pressure and a lower concentration of chlorine in Formula XII.

Lubricants for Pharmaceutical Pill and Tablet Manufacture

These products must be nonflammable and leave only a Food Grade [(Generally Recognized As Safe (GRAS)-listed] residue on surfaces contacted by the pharmaceutical item. Following is a typical formula:

XIII. 2.0%   Lecithin (Soy Bean Source)
       0.5%   Sorbitan Trioleate
       2.5%   Ethanol - Anhydrous (Pure Grain Spirits)
       70.0% CFC-113 (Especially purified)
       25.0% CFC-12

Since a high-purity, nonflammable, volatile liquid is required, which can currently only be met by CFC-113, a 32% immediate reduction (in ozone depletion potential) can be made by using the following formula:

XIV.	2.0%	Lecithin (Soy Bean Source)
	0.5%	Sorbitan Trioleate
	2.5%	Ethanol - Anhydrous
	65.0%	CFC-113
	10.0%	HCFC-142b
	20.0%	HCFC-22

If and when HCFC-123 becomes commercially available, the formulation could then be revised to the following:

XV.	2.0%	Lecithin (Soy Bean Source)
	0.5%	Sorbitan Trioleate
	2.5%	Ethanol - Anhydrous (Pure Grain Spirits)
	77.0%	HCFC-123
	18.0%	HCFC-22

If HCFC-123 does not become commercially available, the following alternative nonflammable formula can be offered:

XVI.	2.0%	Lecithin (Soy Bean Source)
	0.5%	Sorbitan Trioleate
	2.5%	Ethanol - Anhydrous (Pure Grain Spirits)
	55.0%	HCFC-141b
	30.0%	HCFC-124
	10.0%	HCFC-22

Solvent-Cleaners, Dusters, and Coatings for Electric/Electronic Equipment

The solvent-cleaners and dusters for delicate electric and electronic items consist almost uniformly of the following:

XVII.	75%	CFC-113
	25%	CFC-12

Except for the inclusion of approximately 1 to 3% active ingredients, various coating sprays are similarly formulated.

CFC-113 is unique because of its nonflammability, relative volatility, high purity, compatibility with materials of circuit-board and other construc- tions, and for its selective solvent action on greases, oils, and solder by-

products. In some instances, combinations of other solvents may suffice, such
as 1,1,1-trichloroethane, followed by deionized water and then heat lamps.
However, the best immediate course of action may be to replace the CFC-12 (and
perhaps a part of the CFC-113) with other propellants, as shown below:

XVIII.	82%	CFC-113
	18%	HCFC-22
	or	

XIX.	70%	CFC-113
	16%	HCFC-142b
	14%	HCFC-22

If HCFC-123 becomes commercially available, this nonflammable liquid
could be formulated as follows:

| XX. | 83% | HCFC-123 |
| | 17% | HCFC-22 |

Alternatively, the following replacement formula can be offered:

XXI.	60%	HCFC-141b
	30%	HCFC-123
	10%	HCFC-22

A final alternative takes advantage of the high-pressure azeotrope of
HCFC-22/Propane (68:32), which has a pressure of about 135 psig at 70°F and
therefore cannot be added to aerosol cans except as individual ingredients.
The blend is flammable, but it can be rendered nonflammable by the use of
HCFC-123; for example:

XXII.	87.00%	HCFC-123
	4.16%	Propane A-108
	8.84%	HCFC-22

Metered-Dose Oral and Nasal Inhalation Pharmaceutical Products

This category now accounts for about 4 million pounds of CFCs a year in the U.S. (15% of all CFC aerosol uses) and includes approximately 70% of the 152 million CFC aerosol product units sold annually. The dollar value of these products is very high, allowing latitude in reformulating them with more costly propellants. The average canister holds 16.0 g, or about 0.56 Av.oz. The estimated formulas of five high-volume metered-dose inhalant drug (MDID) products are shown in Table 7.

The ultimate conversion of the various "metered-dose inhalant drugs" (MDIDs) to non-CFC formulations represents the greatest individual challenge of all aerosol changeovers. Industry contacts repeatedly assert that MDID is a life-preserver for thousands of dependent users.

Of the presently available "CFC alternative propellants" only HCFC-22 is nonflammable. The nonflammable attribute is highly desired because most MDID users are heavy smokers and may be holding cigarettes while using the product. (The importance of this aspect is arguable.) However, other problems with HCFC-22 are listed below:

a. Teratogenicity

   •   HCFC-22 inhaled at 50,000 ppm for 40 hours per week by preg-
       nant rats in the middle trimester resulted in 0.15% anophthal-
       mia and 0.05% cryptophthalmia; Imperial Chemical Industries,
       Ltd. (ICI) considers these findings to be statistically
       significant.

b. Other Toxicological Aspects

   •   Inhalation toxicology studies beyond those now completed would
       be required for industry approval. The FDA is satisfied with
       the present array of results.

TABLE 7.   ESTIMATED FORMULAS OF FIVE MDID PRODUCTS (PERCENT)

Ingredients	Broncho-dilator	Broncho-dilator	Broncho-dilator	Steroid	Steroid
Drug	1.00	0.147	0.714	0.24	0.386
Sorbitan Trioleate	1.00	0.853	1.000	--	1.000
Ethanol - Absolute	--	--	--	1.00	--
CFC-11	20.00	15.000	24.571	--	20.000
CFC-12	78.00	69.000	49.144	98.76	58.614
CFC-114	--	15.000	24.571	--	20.000

- The drugs now used for adrenergic bronchodilator and corticosteroid therapy are themselves teratogenic to animals, and
  there is a concern that one teratogen (HCFC-22) could rein-
  force the teratogenic activity of another (the drugs). Thus,
  teratogenic testing of any proposed new-propellant products
  would be needed.

- The greater solvency--thus mucosal and tissue permeation--of
  HCFC-22 could cause a more concentrated drug tide to be
  carried through the pre-alveolar tubes, leading to possible
  toxicological and "toxic shock" effects.

c.   Reduced Liquid-phase Density

- A predominantly HCFC-22 product would be about 10 to 15% lower
  in density than the usual MDID items; e.g., 1.2 g/mL, compared
  with 1.4 g/mL. This could cause faster settling of the solid
  drug in the powder dispersion formulas that make up about 90%
  of the industry sales volume.

- Since the meter-spray valve is volumetrically based, about 10
  to 15% higher percent by weight levels of the drug and
  excipients would be required. Likewise, labeled net weight
  would have to be reduced or converted to labeled net volume.

d.   Increased Product Pressure

- The pressure of HCFC-22 is 297 psig at 130°F; whereas, that of
  CFC-12 is 180 psig at 130°F. Under Title 49 of the Code of
  Federal Regulations (CFR), the Department of Transportation
  regulates the interstate shipment of most aerosols [Sections
  173.306(a) and 173.1200(8)]. Most aerosols are not permitted
  to have pressures exceeding 180 psig at 130°F, and containers
  for these products are constructed accordingly. (Applies to
  containers larger than 118.2 mL capacity.)

- Unless the HCFC-22 pressure is significantly reduced by the use of suppressant type propellants (such as HCFC-142b), enhanced leakage rates and other deterrent factors might develop.

e. Increased Solvency

- HCFC-22 is a substantially better solvent than CFC-11, CFC-12 or CFC-114, and could dissolve or partly dissolve certain micronized microcrystalline drugs. In the latter case, unwanted crystal growth could occur as the larger crystals grow at the expense of the smaller ones. This could eventually reduce delivery of the product to the sub-cilial pulmonary regions or even cause meter-spray valve blockage.

f. Difficulties in Reformulation

- HCFC-22 has been reported by 3M (Riker Laboratory) and other drug houses to complicate product redevelopment.

A better approach might be formulas such as the following:

XXIII.
Drug	0.50%
Sorbitan Trioleate	1.00%
CFC-11	13.50%
HCFC-142b	51.00%
HCFC-22	34.00%
Pressure:	61 psig 70°F (air-free)
Flammability:	Nonflammable*
Density:	1.19 g/mL at 70°F

*Both the slurrying liquid and the propellant blend are non-flammable in air.

or one based on HFC-134a as the main propellant, such as the following:

XXIV.   Drug   0.50%
         Sorbitan Trioleate   1.00%
         HCFC-141b   9.50%
         CFC-11   4.00%
         HFC-134a   75.00%
         HCFC-124   10.00%

         Est. Pressure: 70.5 psig at 70°F (air free)
              *   180 psig at 130°F (air free)
         Flammability   Nonflammable*
         Est. Density: 1.20 g/mL at 70°F

         ____
         *Both the slurrying liquid and the propellant blend are non-
         flammable in air.

Following are several reasons to develop a formula based on HFC-134a as
the main propellant:

a.   HFC-134a looks extremely promising as a refrigeration/air con-
     ditioning fluid and aerosol propellant of exceptionally low toxic-
     ity. It also has no Cl or Br atoms, and therefore does not affect
     the stratospheric ozone layer.

b.   HFC-134a may eventually be produced in numerous countries for
     refrigeration/air conditioning uses, so that worldwide availability
     should not be a problem.

c.   The pressure of the system is comparable to the pressures of the
     current MDIDs; e.g., 70.5 psig at 70°F compares to 52 to 70.7 psig
     at 70°F (air free).

d.   The solvent activity of HFC-134a is comparable to that of the CFC-
     12 now in use for MDIDs.

e.   The density of Formula XXIV, while only about 85% that of the usual
     MDID should not pose significant formulation problems.

To maintain the same liquid volume (and canister size) and the same
number of doses per package, the drug content would have to be
increased by about 18% (w/w) for the same meter-spray dosage
volume, and the net weight would have to be decreased by 15%.

Short of retaining current CFC formulas, there are no options for
formulating MDID products at their current density levels of 1.34
to 1.40 g/mL at 70°F, except, possibly, for using Freon C-318
(perfluorocyclobutane: C_4F_8)--a low-pressure food-grade propellant
commercialized during the 1970s but now discontinued because of its
high cost, absence of market, and other factors. DuPont has been
urged to re-examine the merits of this propellant, which has the
desirable property of very low solvency.

f. The highest-pressure propellant is present in the largest per-
    centage. This is equivalent to the present MDID formulas, where
    the percentage of CFC-12 present is from about 50 to 98. It allows
    the product to tolerate minor gas seepages (micro-leakage), which
    occur in all aerosols, but they have more significant effects on
    small dispensers with limited amounts of propellant.

Other important Formula XXIV considerations are presented below:

1. HCFC-141b (very slightly flammable; B.P. = 90°F) is very low in
    toxicity (in testing done to date) and is the primary slurrying
    agent.

    Because of the slight flammability, it is necessary to add not more
    than 4.00% CFC-11 (B.P. = 74°F) to produce a nonflammable solution.
    This approach has the following problems:

    •   CFC-11 is a controlled stratospheric ozone depletion agent.

- It would complicate the NDA process if ultimate reformulation and/or different processing techniques become appropriate in the future. (Two NDA "openings" are more costly than one.)

Alternatives to using CFC-11 as part of the slurry follow:

- Use HCFC-123, which is nonflammable (Formula XXV).

- Use a blend of HCFC-123 and HCFC-141b that is shown to be nonflammable and contains the highest practical level of slightly flammable HCFC-141b to minimize pulmonary exposures to HCFC-123 (Formula XXVI).

- Use 100% slightly flammable HCFC-141b as the slurrying agent, taking necessary precautions to minimize the possibility of fire and/or explosion (Formula XXVII).

- Replace CFC-11 with CFC-113 (Formula XXVIII) for the following reasons:

  -- Its relative ozone-depletion potential is 80% that of CFC-11 (9).

  -- CFC-113 is less volatile than CFC-11 and can be handled in drums rather than cylinders.

  -- CFC-113's nonflammability in blends is similar to that of CFC-11, reflecting its lower volatility and higher molecular weight.

2. The "true propellant" portion is 75 parts of HFC-134a and 10 parts HCFC-124 (Formulas XXIV through XXVII). The latter is added to reduce the vapor pressure of the HFC-134a into the usual "DOT Specification 2Q" pressure range of less than 180 psig at 130°F. This means that the usual aluminum or stainless steel canisters will

suffice. However, for containers of less than 4 fluid ounces
(118.2 mL), this is not a requirement.

Using straight HFC-134a is another option, if the extra pressure
can be accommodated. The additional pressure would be about 5 psi
at 70°F and 12 psi at 130°F. The consequences follow:

- A statistical reduction of HCFC ozone depletion by removing
  the HCFC-124 (Formula XXIX).

- Reduction in formula complexity by eliminating one ingredient
  (Formula XXX).

- Review of pressure-resistant qualities of the presently used
  aluminum canisters and vials. The supplier would have to
  guarantee their safe use for products with vapor pressures to
  about 192 psig (130°F) air free, and/or about 210 psig (130°F)
  if atmospherically clinched.

- Review of the Department of Transportation (DOT) position.
  Two pages from the present tariff containing regulations for
  compressed gases in dispensers of 7.5 cubic inches (4.0 fluid
  ounces, or 118.3 ml) or less appear in Appendix B of this
  report. Pressure limitations are not described. Products
  pressurized with propane (unsuppressed) are known to be sold
  in standard aluminum containers for nail guns and other uses.
  As currently produced, they have pressures of about 123 psig
  at 70°F and 278 psig at 130°F--with atmospheric crimps.

Table 8 summarizes the above detailed discussion of the various formulas, compares their properties, and gives preference ratings. (The pressure
data in Table 8 are on an air-free basis. For 20″Hg° vacuum crimping, add 7
psi at 130°; for atmospheric crimping add 22 psi at 130°F.) The preference
ratings in Table 8 assume that pharmaceutical firms and their fillers cannot
reasonably handle the filling of slurries made with flammable dilutents having

TABLE 8. COMPARISON OF RECOMMENDED MDID FORMULAS

Formula Number:	XXIV.	XXV.	XXVI.	XXVII.	XXVIII.	XXIX.	XXX.
Ingredients (%)							
Drug	0.5	0.5	0.5	0.5	0.5	0.5	0.5
Excipient(s)	1.0	1.0	1.0	1.0	1.0	1.0	1.0
HCFC-123		13.5	6.5				6.5
HCFC-141b	9.5		7.0	5.3	9.0	9.5	7.0
CFC-11	4.0			1.5		4.0	
CFC-113					4.5		
HFC-134a	75.0	75.0	75.0	66.7	75.0	85.0	85.0
HCFC-124	10.0	10.0	10.0	25.0	10.0		
Estimated Properties							
Pressure (70°F psig)	71	71	71	69	71	76	76
Pressure (130°F psig)	180	180	180	175	180	192	192
Density g/mL 70°	1.20	1.19	1.19	1.21	1.20	1.18	1.17
Slurry Flammable?	No	No	Border	Border	No	No	Border
Strat. O_3 Depletion Potential (CFC-11 = 1.00)	0.046	0.007	0.007	0.024	0.042	0.043	0.004

Preference Ratings:

Order of Preference (1 is Highest; 7 is Lowest)	If Toxicity Rating of HCFC-123 is Equal to or Better Than That of CFC-11	If Toxicity Rating of HCFC-123 is Significantly Worse Than That of CFC-11
1	XXV.	XXVII.
2	XXX.	XXIX.
3	XXVI.	XXIV.
4	XXVII.	XXX.
5	XXIX.	XXVI.
6	XXIV.	XXVIII.
7	XXVIII.	XXV.

flash points below room temperature. Otherwise, slurries of predominantly HCFC-141b would be recommended.

Finally, it is recommended that the MDID development chemists concentrate in two areas, as illustrated by the formulas shown in Table 9.

HCFC-124 is preferred over presently available HCFC-142b, because the latter is slightly flammable and the ability of HFC-134a to suppress this flammability is unknown.

When calculating the stratospheric ozone depletion values in Table 8, in the absence of complete data, the relative depletion potential of the various HCFC materials was calculated as 0.03 (based on CFC-11 = 1.00). The ozone depletion values of HCFC-22 and HCFC-142b are about 0.05 and 0.03, respectively. Using the 4.0 million-pound domestic use level (1988) of CFC propellants as a basis, Table 10 shows the calculated ozone depletion levels of the two MDID formulas ranked highest.

About 10% of the MDID aerosol volume consists of aerosols in which the drug item is in solution instead of in the usual microcrystalline suspension. In these formulas, if the drug can be dissolved in the concentrate (such as ethanol) and all halogenated ingredients added as a nonflammable propellant, then formulation possibilities become easier. The transition is suggested by the formulas shown in Table 11.

TABLE 9.   FINAL RECOMMENDED MDID FORMULA PROTOTYPE

Formula No. Ingredients (%)	XXV. Assuming Toxicity Rating of HCFC-123 is as Good as or Better Than That of CFC-11	XXVII. Assuming Toxicity Rating of HCFC-123 is Adequate, But Significantly Worse Than That of CFC-11
Drug (Microcrystalline Suspension)	0.5	0.5
Excipient(s) (As Sorbitan Esters)	1.0	1.0
HCFC-123	13.5	
CFC-113		4.5
HCFC-141b		9.0
HFC-134a Propellant	75.0-85.0	75.0-85.0
HCFC-124 Propellant	10.0-none	10.0-none

TABLE 10.   CALCULATED OZONE DEPLETION FROM MDID ITEMS
(POUND EQUIVALENCE OF CFC-11)

Year	1988	1993
Present Formulas (No changes)	4.000 (MM lbs)	5.000 (MM lbs)
Formula XXV		
With 10% HCFC-124 (0.007)[a]	0.028 "	0.035 "
Without HCFC-124 (0.004)[a]	0.016 "	0.020 "
Formula XXVII		
With 10% HCFC-124 (0.042)[a]	0.168 "	0.210 "
Without HCFC-124 (0.039)[a]	0.156 "	0.195 "

[a]Formula's overall ozone depletion potential.

TABLE 11. PROPOSED FORMULA TRANSITION FOR SOLUBLE MDIDS

Ingredients	Present Formula, %	Proposed Formula, %
Soluble Drug	0.5	0.5
Sorbitan Trioleate	1.0	1.0
Ethanol - Absolute	8.5	8.5
CFC-114	36.0	
CFC-12	54.0	
HFC-134a		54.0
HCFC-124		36.0
Pressure (psig at 70°F, air free)	52.	66.
Stratospheric O_3 Depletion	0.88	0.01
Flammability		
Concentrate Solution	Flammable	Flammable
Total Product	Nonflammable	Nonflammable

Note: Ethanol (Absolute) has an Open Cup Flash Point of 65°F and is readily handled and filled in explosion-resistant settings.

Contraceptive Vaginal Foams

The present-day products of this sub-category are typified by the following formula:

XXXI.   Contraceptive (Spermicidal) Drug   10.5%
               Triethanolamine Myristate/Laurate   0.5%
               Deriphat 151C (Amphoteric or
                 "Zwitterion" Surfactant)   4.0%
               Deionized Water   77.0%
               CFC-114   3.2%
               CFC-12   4.8%
               Pressure (psig at 70°F, air free)   47.
               Density (g/mL at 70°F)   1.031
               Flammability   Nonflammable

A reasonable replacement formula is as follows:

XXXII.   Above concentrate   92.0%
               Additional Deionized Water   4.0%
               Propellant A-46   4.0%
                 16% Propane A-108
                 84% Isobutane A-31

               Pressure (psig at 70°F, air-free)   46.
               Density (g/mL at 70°F)   0.983
               Flammability   Transient*

               _____

               *If the foam puff is touched with a lighted match, a blue
               flame will leap across the surface and immediately go out.
               The process can be repeated at will. Otherwise, the product
               is nonflammable because of the 91.0% water content.

If even this vestige of flammability cannot be tolerated, or if a more quick-breaking type of foam is wanted, at least 25% of the propellant must be converted to HFC-152a (CH_3-CHF_2), as in Formula XXXIII:

XXXIII.   Above concentrate   92.0%

Above concentrate	92.0%
Additional Deionized Water	3.5%
HFC-152a	1.5%
Propellant A-60	3.0%
33% Propane A-108	
67% Isobutane A-31	
Pressure (psig at 70°F, air-free)	61.
Density (g/mL at 70°F)	0.987
Flammability	None**

      **By the usual test procedures. However, the propellants
      (individually or pre-blended) are flammable and must be
      dealt with under highly controlled conditions.

The FDA (Drug Division) has jurisdiction over this product via their NDA program. They may require the use of a nonflammable propellant.

Apart from nitrous oxide (N_2O), which would cause a meter-foam valve to deliver significantly more drug near the end of the can than at the beginning, the only nonflammable, non-CFC choice would be certain blends of HCFC-22 (at least 40%) and HCFC-142b. The formula is as follows:

XXXIV.

Above concentrate	92.0%
Additional Deionized Water	1.5%
HCFC-22	2.6%
HCFC-142b	3.9%
Pressure (psig at 70°F; air free)	64.
Density (g/mL at 70°F)	1.013
Flammability	None

    NOTE: The pH value must not exceed 8.2 at 77°F, or the
                HCFC-22 will hydrolyse, reducing the pH, and forming
                chloride ion that may act to corrode the aluminum
                canister.

Various possibilities for reformulation also exist for the "future alternative" propellants, such as a blend of HFC-134a/HCFC-142b or HFC-134a/HCFC-124, but these may be 7 to 10 years away from commercialization.

Formula XXXII may be the best alternative, since hydrocarbon propellants are highly flammable and must be handled and filled in well-ventilated enclosures containing only explosion-proof equipment and outfitted with an appropriate electroprotective system for detection, alarm, and other automatic responses.

Mercaptan (Thiol) Warning Devices

Present formulations of this product are essentially like the following:

XXXV.   Ethyl Mercaptan (CH_3-CH_2-SH)   2.0%

             CFC-12   98.0%

This product is used rarely, but can be employed to detect methane and carbon monoxide gases in coal mines and to automatically release the stench to warn those unable to hear the audio-alarms. The released material must be nonflammable.

An alternative formula, using presently available propellants is as follows:

XXXVI.   Ethyl Mercaptan   2.21%
             HCFC-22   39.12%
             HCFC-142b   58.67%

             Pressure (psig at 70°F; air free)   64.5
             Density (g/mL at 70°F)   1.217
             Flammability: None (but borderline)

Fillers who do not have suitable facilities for handling flammable propellants must purchase the HCFC-22/HCFC-142b (40:60) nonflammable blend.

Intruder Alarm Devices

Present formulations consist of the following:

XXXVII.   100.0%   CFC-12

These 6 and 12 Av.oz. cans are used as home, car, and boat sensory alarm equipment to sound a high-intensity (90-110 decibel) horn if the system detects movement, selected sounds, or other occurrences. Since large amounts of propellant are emitted under conditions beyond the marketer's control, e.g., in computer rooms, near sparking equipment, etc., it is necessary to use a nonflammable composition or an electric alternative device.

Most of these dispensers use an internal dip tube to carry the liquified gas through the valve and into the expansion chamber of the horn, where it expands to the gaseous state. One mL of CFC-12 produces 256 mL of gas at 70°F and atmospheric pressure. Strong chilling of the chamber and icing of the humidity in the air may occur, but this has no effect on the horn properties. Nevertheless, to relieve such icing (but not the chilling), some formulas are revised to the following:

XXXVIII.   85.0%   CFC-12
            15.0%   Ethanol Absolute (or Methanol Absolute)

In such cases, 1 mL of product produces 236 mL of CFC-12 and alcohol vapor in use. The mixture is still nonflammable.

Conversion to non-CFC ingredients can be readily effected using the following composition:

XXXIX.   60.0%   HCFC-142b
          40.0%   HCFC-22

          Pressure: 63 psig at 70°F (air free)
          Pressure: 69 psig at 70°F (22"Hg° vacuum crimp)

Since this composition is just barely nonflammable, there is no latitude for adding (flammable) alcohols for de-icing purposes. Some conversions to 100% HCFC-22 in heavier cans have already taken place.

Also, unlike pure CFC-12, a distillation effect will accompany any additions of strongly chilled Formula XXXIX to an empty can. The preferential evaporation of nonflammable HCFC-22 will cause the residual liquid in the can to become technically flammable and lower in pressure. Although the flammability will be of small consequence unless the HCFC-22 sinks to about 32.0% or so, these effects do constitute a drawback. Any fillers using refrigeration-filling methods will have to change to pressure-filling techniques, including the drawing of a partial vacuum (22"Hg° minimum) in the can before introducing the liquid to avoid this effect. The density will be 15% lower than for CFC-12, but the gas volume per mL will be higher.

Flying Insect Sprays for Food-Handling Areas

The present exemption is for non meter-spray and non total-release spray pesticides in commercial food-handling areas.

A typical formulation for such a product would be as follows:

XL.
2.0%	Toxicants
13.0%	Anhydrous Ethanol (As S.D. Ethanol 40-2; 200°)
42.5%	CFC-11
42.5%	CFC-12

For the following reasons, most of the original product volume has been replaced with water-based formulas when using hydrocarbon gases as the propellant:

- The spray cannot be ignited with a flame because of the presence of water;

- Food-handlers were using the CFC-type sprays for locker rooms, bathrooms, offices, and other areas in the establishment; and

- The factory cost of a 16 Av.oz. flying insect killer, using CFCs, was $0.42/can higher than the cost of the water-based, hydrocarbon product, not considering manufacturing losses, which would add $0.03 to $0.05/can. This would increase to about $1.00/can more (minimum) at wholesale prices.

A typical hydrocarbon product formula is as follows:

XLI.   2.0%   Toxicants
        1.0%   Petroleum Distillates (Food Grade)
        0.4%   Non-ionic Emulsifier (As Tween 80, by ICI America)
        0.1%   Morpholine
        0.1%   Sodium Benzoate (Food Grade)
        64.4%   Deionized Water; USP
        30.0%   Hydrocarbon Blend BIP-40 (Food Grade)
                  18% Propane A-108
                  33% Iso-butane A-31
                  49% n-Butane A-17

Pressure: 47 psig at 70°F (21"Hg° vacuum crimp)
Delivery Rate: 0.62 g/sec at 70°F - initial
Density: 0.904 g/mL at 70°F
Flammability: None, by standard tests

NOTE: The hydrocarbon blend is extremely flammable, and
       special methods must be employed to handle and fill
       it safely.

If an anhydrous, nonflammable formulation is regarded as "essential,"
the following would suffice:

XLII.   2.0%   Toxicants
        42.0%   HCFC-22
        56.0%   HCFC-142b

Pressure: 69 psig at 70°F (with 22"Hg° vacuum crimp and
           gas purge)

Density: 1.213 g/mL at 70°F.

The cost, however, may preclude serious interest in this formulation.

Flying Insect Sprays for Aircraft

The present fumigation sprays for aircraft cabins are composed of the
following:

XLIII.   1.0%   Pyrethrins/Cinerins (20% in Petroleum Distillates)
              1.0%   Piperonyl Butoxide - Technical
              49.0%   CFC-11
              49.0%   CFC-12

Other toxicant systems, such as 2.0% Sumethrin or 2% Resumethrin, are beginning to be used because of the high-cost and sporadic unavailability of Pyrethrin/Cinerin natural products. Products with the same formulation as Formula XL above are used in some areas.

The most reasonable alternative would be the following:

XLIV.   2.0%   Toxicants (As described just above) .
              40.0%   HCFC-22
              58.0%   1,1,1-Trichloroethane - Inhibited (4).

              Pressure: 68 psig at 70°F (with 22"Hg° vacuum crimp and
                         gas purge)

              Density: 1.270 g/mL at 70°F
              Flammability: None

If the generally acceptable odor of 1,1,1,-trichloroethane-inhibited is too noticeable or disagreeable, the much more costly Formula XLII could be substituted.

Flying Insect Sprays for Tobacco Barns

As discussed in Section 3, these sprays consist entirely or almost entirely of 9-, 15- and 25-pound refillable cylinders containing 40 to 60% "Vapona" (organo-phosphate) insecticide and 60 to 40% CFC-12. These containers are not classified as "aerosols" by the aerosol industry. Nevertheless, it is appropriate to identify an acceptable alternative composition, such as the following:

XLV.   40 to 65% Vapona™ Insecticide - Technical (95% min.)
              35 to 60% HCFC-22

         Pressure: About 75-100 psig at 70°F (air-free refill)
         Density: About 3 to 7% lighter than the CFC-12 product

         NOTE: HCFC-22 is a better dispersant than CFC-12, allowing
                 the use of a somewhat higher level of Vapona (65%
                 maximum). For such formulas, the standard amount of
                 Vapona per cylinder can be used, and the same
                 coverage per refill will apply.

This formula range is too high in pressure for "aerosol" containers, but
they can be readily replaced (if used) with cylinder forms for this application.

Aircraft Maintenance and Operation Sprays

These sprays may be either nonflammable (CFC-based) or flammable (hydro-
carbon-based) according to the conditions of use. For example, exterior
touch-up paint could be flammable, typically using 29% Hydrocarbon Blend A-70
to A-85.

The nonflammable types include dusters, lubricants, and mold/mildewcides
to be used in such areas as the cockpit, engine, and radio/radar area.
Otherwise, air pockets within instruments and between close-packed equipment
could be over-sprayed into the flammable range and such vapors be ignited by a
spark source.

This means that, whenever possible, both the propellant and the total
product should be nonflammable. Also, the use of solvents such as 1,1,1-
trichloroethane-inhibited would often be contraindicated because of unwanted
effects on the delicate electrical or electronic gear.

A typical alternative formulation could be patterned after Formula XII.
A typical formulation for a mold/mildewcide would be as follows:

XLVI.   0.12%   o-Phenylphenol (As Dowicide 1 - Dow Chemical)
           0.38%   N-Morpholinium Soya Ethosulfates (G-271 - ICI)
           0.10%   Morpholine
           0.10%   Quaternary Ammonium Inhibitor (Q.A.I.-nitrite)
          24.30%   Deionized Water
     (Negligible   Acid form of Deriphat or Miranol Amphoteric
        Amount)   "Zwitterion" Surfactants to pH = 8.0 - 8.2

          50.00%   Ethanol (Anhydrous) As SD Alcohol 40-2 200°
          25.00%   HCFC-22

          NOTE: The water content may have to be reduced slightly,
                    depending on pressure and phase compatibility.
                    Pressure should not exceed 75 psig at 70°F.

                    The Set-A-Flash Open Cup flash point should not be
                    below about 100°F for this formula because of the
                    presence of water.

If even this minor degree of product flammability is of concern, a reformulation to use future alternative propellants such as the following will be necessary:

XLVII.   0.20%   BTC-2125M (80% A.I. quaternaries) Stepan Chem.
           0.10%   Quaternary Ammonium Inhibitor (Q.A.I.-nitrite)
          59.70%   HCFC-141b
          40.00%   HCFC-124

          Pressure: 32 psig at 70°F (22"Hg° vacuum-crimp)
          Density: About 1.28 g/mL at 70°F
          Flammability: None

          NOTE: The slight flammability of HCFC-141b is countered by
                    HCFC-124.

Military Aerosols

As mentioned in Section 3, the bulk of these products are insecticides with the following formulation:

XLVIII.   2.0%   Toxicants
          49.0%   CFC-11
          49.0%   CFC-12

The military's rationale for requiring a nonflammable formulation is that they have no control over where these dispensers are used. For example, if a can were used to kill a roach in the cockpit of a fighter-jet, a sparking source could start a fire in a very confined, high-technology environment.

Partial resolution of this problem requires use of a nonflammable blend of presently available (ideally) HCFC propellants, e.g.:

XLII.   2.0%   Toxicants
            42.0%   CFC-11
            56.0%   HCFC-142b

For other "critical area," nonflammable military-use formulations, refer to the descriptions of the following formulas:

Formula XLVII: Mold/Mildewcide;

Formula XII: Lubricant; or

Formulas XX or XXI: Duster.

Diamond Grit Sprays

If this product type still survives in CFC-form, and if the amount of concentrate is 5% or less (flammable or nonflammable), or 20% or less (non-flammable), an alternative formulation to yield a nonflammable aerosol spray product could be the following:

IL.   2 - 20%   Concentrate Liquid
          98 - 80%   HCFC-Blend:
                         58% HCFC-142b
                         42% HCFC-22

CFC-115 for Aeration of Puffed Food Products--Certain Limitations Applied

This product use of CFC-115 has been discontinued because of a Department of Transportation (DOT) "Special Exemption" allowing the introduction of additional nitrous oxide (N_2O) propellant. No rekindling of interest in

CFC-115 for this use is anticipated, although no nonflammable replacement product is commercially available.

The following paragraphs cover excluded products (where the propellant is all or part of the "product."

Tire Inflators

Only one marketer is thought to be using CFCs for this product type. The approximate formula is as follows:

L.   45%   Water-based Sealant Concentrate
            55%   CFC-Blend:
                      40% CFC-114
                      60% CFC-12

A number of deaths have occurred during the repair of tires containing an explosive mixture of air and hydrocarbon propellants, and insurance firms sometimes refuse to continue product liability coverage. The above formula has been one result of this. It is more costly and less effective than the standard 25 to 30% Hydrocarbon Blend A-46 formulas.

An alternative nonflammable formula would be as follows:

LI.   65%   Water-based Sealant Concentrate
            23%   HCFC-142b
            12%   HCFC-22

            Pressure: 56 psig at 70°F (air free)
            Density: 1.087 g/mL at 70°F
            Flammability: None--but borderline.

If a lower pressure is required, formulations would have to use future alternative HFC-134a in place of the HCFC-22.

Polyurethane Foams

The typically 30-35% CFC-12 used in these formulas as a nonflammable blowing agent consumes about 3.1 million pounds per year of CFC-12. The alternative agent must be nonflammable and chemically compatible with the foam. The following formulation is suggested:

LII.   66%   Polyurethane pre-polymer dispersion
                2%   Dimethyl Ether (DME) (Moisture Scavenger)
               17%   HCFC-142b
               15%   HCFC-22

        Pressure: 48 psig at 70°F (air free)
        Flammability: Combustible, because of the concentrate.
              The HCFC blend is nonflammable.
              The HCFC/DME blend is marginally nonflammable.

The pressure is about the same as is exerted by the CFC-12 in present formulas. The vapor volume would be about 26% greater. If less vapor volume is desired, 14% HCFC-142b and 12% HCFC-22 could be used. The DME ties up unwanted moisture in the aerosol can and prolongs the service life of the product.

Chewing Gum Removers

The current formula is 100% CFC-12, and the product is designed as a simple chiller for embrittlement of the gum, after which the frangible mass can be cracked apart and removed. No solvents can be used, or the chilling will be mitigated and softening of the gum surface may result.

The proposed alternative formula is as follows:

LIII.   44.0%   HCFC-22
           54.0%   HCFC-142b

        Pressure: 70 psig at 70°F (22"Hg° v.c. plus purge)
        Density: 1.24 g/mL at 70°F (CFC-12 is 1.34 g/mL 70°F)
        Flammability: Nonflammable

Drain Openers

This product was developed by Glamorene, Inc. and marketed as "Drain Opener" by them. It was later sold to Lehn & Fink Products Group (Sterling Drug Co.) and then discontinued because of the high price and numerous in-use problems.

A formula like Formula No. LIII could be proposed for a non-CFC version, but the item is now dead, and no marketers seem to be interested in it as a future product possibility.

Chillers - Medical

These are presently CFC-12 or CFC-12/114 blends for topical application before localized incisions are made for removing warts, minor birthmarks, etc. Following is a suitable, currently available replacement formula:

LIV.   10.0%   Ethanol - Absolute (Pure Grain Spirits)
          50.0%   HCFC-142b
          40.0%   HCFC-22

          Pressure: 67 psig at 70°F (22"Hg° vacuum crimp)
          Density: 1.19 g/mL at 70°F
          Flammability: Nonflammable propellant blend and nonflammable
                         product by standard tests.
                  (The ethanol has a Set-a-Flash Closed Cup = 56°F.)

If desired, the ethanol can be removed from this formula. It is included as a pressure depressant, cost-reducer, and germicidal solvent. Medically-approved specially denatured ethanol grades, such as SDA-17, may be used in some cases.

Boat Horns

Formula IXL (described above) is suggested:

LV.   60.0%   HCFC-142b
          40.0%   HCFC-22

Non-Electric/Electronic Dusters

Products in this class are used to blow dust off photograph records, tape decks, stylus tips, and so forth. If hydrocarbons were used, a flammable condition could result; e.g., 9.7 g of butanes will cause all the air in a 55-gallon drum to become flammable. Therefore, dusters should be nonflammable or essentially nonflammable in composition. The inclusion of solvents is contraindicated in general.

Formula XXXIX, discussed previously, is suggested:

XXXIX.   60.0%   HCFC-142b
           40.0%   HCFC-22

           Pressure: 63 psig at 70°F (air free)
           Pressure: 69 psig at 70°F (22"Hg° vacuum crimp)

While these propellants are better solvents than CFC-12, their residence time on target surfaces would probably be too brief for any adverse effect.

Microscope Slide Cleaners and Related Products

When the target surface is not adversely affected by solvents that are more active than CFC-113, and when slightly slower evaporation rates can be tolerated, the following formula is recommended.

LVI.   72%   1,1,1-Trichloroethane - Inhibited (4)
           28%   HCFC-22

           Pressure: Estimated as about 48 psig at 70°F
                      (with vacuum crimp)
           Density: 1.302 g/mL at 70°F
           Flammability: Nonflammable

When extreme purity and the evaporation rate and other properties indigenous to CFC-113 are required, the following formula can be used while research is conducted about the possibility of replacing the CFC-113 with HCFC-123, or with a blend of about 50% HCFC-123 and 50% HCFC-141b:

LVII.   73%   CFC-113
               27%   HCFC-22

The supplier of these future alternative products would have to be contacted about the possibility of making them available in highly purified forms.

Solvents - Medical

This is a minor use of CFC-113, for which it is uniquely qualified. The Dow-Corning Corporation, a maker of two such products, has examined the possibilities of using 1,1,1-trichloroethane (4) and methylene chloride and has pronounced them unsatisfactory.

Only two other possibilities can be offered:

Adhesive Spray:

LVIII.   3%   Silicone-based Adhesive
               80%   HCFC-141b (Marginally flammable; B.P. = 90°)
               17%   HCFC-22 (max.)

LIX.   3%   Silicone-based Adhesive
               41%   HCFC-141b (Marginally flammable; B.P. = 90°)
               40%   HCFC-123 (Nonflammable; B.P. = 82°F)
               16%   HCFC-22 (Nonflammable)

               NOTES: The Adhesive is marketed as a 50% Active Ingredient
                      dispersion in CFC-113. The supplier should reformu-
                      late the product to use perchloroethylene, 1,1,1-
                      trichloroethane - inhibited, or HCFC-141b, in order
                      of increasing volatility.

                      The preliminary results of toxicological tests now
                      underway make the use of HCFC-123 uncertain.

There is concern about HCFC-141b because of its marginal flammability. This is probably sufficiently neutralized by the 17% HCFC-22 in Formula LVIII to eliminate any intrinsic problems. However, if nonflammability is an absolute requirement, the addition of the HCFC-123 in Formula LIX provides extra confidence, although testing would be required.

Adhesive Solvent Spray:

LX.	84%	HCFC-141b (Marginally flammable; B.P. = 90°)
	16%	HCFC-22 (Nonflammable) maximum
LXI.	42%	HCFC-141b
	42%	HCFC-123
	16%	HCFC-22

NOTES: As above, for HCFC components.

Contingency Products, Including Unauthorized Uses

Most of these will vanish as CFCs become progressively less available, beginning with the 15 to 25% effective reduction in production volume on July 1, 1989. (This effective reduction is due to the growth in CFC use since 1986, the baseline year for the Montreal Protocol-required CFC cutbacks.)

In general, CFC-12 can be replaced with 42% HCFC-22 and 58% HCFC-142b for those possible uses of CFC-12 not covered in this section.

SUMMARY

Of the 26 product categories examined in this section, most can be reformulated to eliminate or contain much lower concentrations of CFCs. The metered-dose inhalant drug (MDID) aerosols are the most difficult. To date, relatively little effort to reformulate products seems to have been expended by any of the CFC users, and this is particularly true of the "future alternatives," perhaps because of toxicological uncertainties and because the alternatives may not become commercially available for several years.

Pharmaceutical marketers (MDIDs, contraceptive foams, etc.) have expressed grave concern about dealing with the FDA Drug Division and reopening their NDA files, especially if numerous firms apply for amended NDAs concurrently, thus overloading the small FDA staff in this research area.

It is technically possible to reduce aerosol-related CFC uses to less than 25% of the present 25.5 million pound volume within five years with existing chemicals. Post-1994 reductions, however, depend on the toxicology of the alternatives and on what possible sacrifices in quality and convenience should be made to accommodate the need for maximum CFC reductions.

5. Procedures for and Costs of Substituting Alternative Formulations for CFC Aerosols

INTRODUCTION

In Section 4, currently available substitutes for CFC aerosols and those that may be available in the future were considered for a total of 26 categories of aerosol products that currently use CFC propellants in whole or in part. Sixty-one formulations were described. For several products, more than one formula transition path was shown, but in each instance a preferred non-CFC formula was recommended.

When substituting alternative aerosol formulas for CFC-based aerosol products, one of the most important considerations is flammability. Depending on circumstances, one or more of three main formulation routes will be taken, as shown in Figure 1.

For firms producing CFC aerosols, the transition to non-CFC formulations will increase in difficulty and cost (and often decrease in acceptability) as the replacement formulas go from TYPE 1A to TYPE 3B (shown in Figure 1).

The costs of CFCs and current alternative propellants shown in Table 12 have been taken from price lists provided primarily by E.I. duPont de Nemours & Co. (Inc.), the only source for all of the gas-liquids. One price increase, effective on February 14, 1989, ranged from no increase for dimethyl ether to 15% increases for CFCs. The HCFC and HFC propellants were intermediate in price range.

Table 12 also includes estimates of the costs of these propellants in 1993 and 1994 and compares them with the present pricing schedule. These costs are speculative and assume a dramatic increase in the price of CFCs.

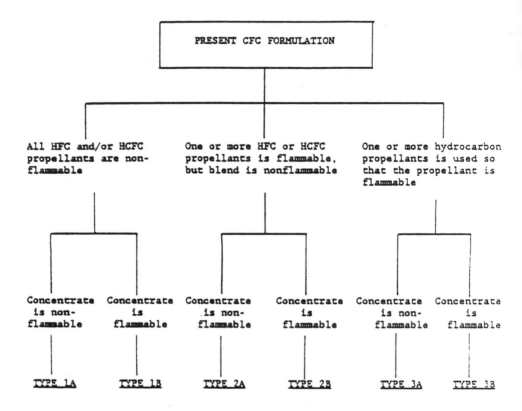

NOTES: A propellant is here defined as "flammable" if it can produce a
flammable composition (or range) in air.

A concentrate is defined as flammable if it exhibits a Set-A-Flash
(Closed Cup) flash point of less than 100°F.

Figure 1.   Aerosol Reformulation Options (1989 - 1993)

TABLE 12.   CURRENT AND 1993/1994 PRICES FOR
VARIOUS AEROSOL PROPELLANTS

Propellant	Current ($/lb)	Estimates for 1993/1994 ($/lb)
CFC-11	0.79	3.00
CFC-12	0.89	3.40
CFC-113	1.05	3.50
CFC-113 (Extreme Purity Grades)	1.40	4.00
CFC-114	1.23	4.0
CFC-115	NA[a]	NA[a]
Dimethyl Ether (DME or Dymel A)	0.38	0.55
HCFC-22	1.05	2.00
HFC-152a	1.60	3.40
HCFC-142b	2.40	3.25
HCFC-123	NA	4.00
HCFC-124	NA	3.70
HFC-134a	NA	4.40
HCFC-141b	NA	3.00
Carbon Dioxide	0.14	0.19
Hydrocarbons (Aerosol Grade)	0.18	0.23

Corollaries:

On a volume basis (gallon versus gallon), CFCs will cost about 42 times as much as hydrocarbons. On a weight basis, they will cost about 15 times as much.

On a volume basis, the HFC and HCFC alternatives will cost about 25 times as much as the hydrocarbons. On a weight basis, they will cost about 12 times as much.

On a volume basis, the future HFC and HCFC alternatives will cost about 38 times as much as the hydrocarbons. On a weight basis, they will cost about 16 times as much. Note that long-term prices are difficult to predict more accurately than within ± 25%.

On a volume basis, dimethyl ether will cost about 3.7 times as much as the hydrocarbons. On a weight basis, it will cost about 2.9 times as much.

[a]Ton cylinders only.

COST OF CONVERTING FILLING LINES

Many of the current fillers and marketers of non-pharmaceutical CFC-type aerosol products are not equipped to handle flammable propellants. They could produce TYPE 1 products with no capitalization, and TYPE 2 products with minor capitalization, but for TYPE 3 products most would have to do one of the following:

- Commit very heavy capitalization, time, and educational resources to convert to hydrocarbon formulas and accept a statistical risk of fires and explosions; or

- Close down their facilities and go to a contract filler able to handle such products.

The cost to convert one medium-speed [80 to 160 cans per minute (cpm)] aerosol line to the filling of TYPE 3 products will vary with safety commitment levels but will range from $400,000 to $1,200,000. A probable average would be $900,000.

Table 5, which appears in Section 3 of this report, lists 34 filler and marketers handling TYPE 1 products. While these firms handle 90% of the non-pharmaceutical CFC business, it is estimated that an additional 30 to 40 firms, often termed "garage" operations, are also filling CFC aerosols. Smaller operators who elected to fill TYPE 3 products would be assuming a substantial risk. They would probably close down their operations or rely on a contract filler for continuing production.

It is estimated that about 12 of the larger CFC fillers would convert one line to filling TYPE 3 replacement formulas, because they currently have no such line and prefer to continue their in-house manufacturing activities for all of their products. Most of the remainder would convert to only TYPES 1 and 2 products or go to a contract filler.

Table 13 shows estimates of the total industry costs of in-house filler conversions from CFCs to the TYPES 1, 2, and 3 products defined in Figure 1.

The installation of a line able to handle TYPE 3 products would also permit the filler/marketer to fill other aerosol products that commonly use flammable products when such in-house filling business is out of reach because of the hazards involved.

The pharmaceutical industry currently has approximately 20 marketers that sell metered-dose inhalation drugs (MDID) and metered-dose contraceptive foams for human consumption under FDA exemptions. At least two U.S. contract fillers serve this group: Armstrong Laboratories, Inc. (West Roxbury, MA) and the 3M Health Care Specialties Division (St. Paul, MN). Approximately 10 production lines are currently engaged in the manufacture of these products.

The approximately 96 million units of microcrystalline MDIDs suspended in the propellant that are manufactured each year could be converted to a TYPE 1A formula (Formulas XXV and XXVII, as described in Section 4) without major cost. These two formulas are recommended.

When the MDID is dissolved in anhydrous ethanol (about 11 million units a year), a TYPE 1B formula (see Table 11) is recommended. Firms handling products in that subcategory are already manufacturing this formula, except for propellant selection; therefore, conversion costs should be minimal.

The last pharmaceutical category is the metered-dose contraceptive foam and its non-metered counterpart. The most reasonable conversion for this product is to a TYPE 3A formula, using hydrocarbon propellant blends. At least some of these products are being manufactured in-house by the marketers. They would have the option of spending about $900,000 per converted line, or going to contract fillers already set up to handle such products.

The above statements do not consider research (redevelopment) costs, regulatory costs (FDA-NDA), or marketing costs--only the manufacturing conversion outlays.

TABLE 13.   COST ESTIMATES FOR PRESENT IN-HOUSE FILLER CONVERSIONS[a]

Conversion Mode	No. of Fillers	Lines	Costs ($)
CFC to TYPE 1	0	0	0
CFC to TYPE 2	12	13	4,000,000
CFC to TYPE 3	12	12	11,000,000
			$15,000,000

[a]Non-pharmaceutical firms only.

METHODS AND COSTS FOR DEVELOPING ALTERNATIVE FORMULATIONS FOR CFC AEROSOLS

The most appropriate way to discuss the cost-effectiveness of alternative approaches is on a product-by-product basis. The basis for the following discussion is 1) use of an average size container, and 2) the information in the "Product Volume of CFC Versions" column of Table 4 in Section 3. The calculated costs of interim formulations are based on an assumed dramatic increase in the price of CFCs.

Mold Release Agents

The per-can cost increase of converting:

3% Concentrate		3% Concentrate
57% CFC-11	to Formula V	65% HCFC-123
40% CFC-12		32% HCFC-22

in the average 1.11 lb dispenser can only be calculated in terms of 1993/1994 propellant prices, because HCFC-123 is not yet available and a price has not been established. Using parity for the concentrate and price estimates from Table 12 the following differential can be calculated:

Diff. = 1.11 x 1.08 x [(65% x $4.00 + 32% x $2.00) - (57% x $3.00 + 40% x $3.40)]

Diff. = 1.20 [$3.24 - $3.07] = $0.204 per can.

Notes:

- Analysis assumes that the current can sizes can hold an extra 15 volume percent of product, since Formula V has a lower density than the present CFC formula (Formula I). Otherwise, a larger can or a smaller net weight will be required, adding to the cost per pound.

- The 1993/1994 prices are obviously speculative.

- The 1.08 factor takes into account an anticipated 8% propellant loss (during filling) for both formulas, which is about the industry average for these products.

Because it may be impractical to wait several years for the commercialization of the HCFC-123 used in Formula V, it is appropriate to look at the immediate conversion option, e.g.,:

Calculate the per-can cost increase of conversion from:

3% Concentrate		3% Concentrate
57% CFC-11	to	53% CFC-11
40% CFC-12		26% HCFC-142b
		18% HCFC-22
TYPE 1A		TYPE 2A

Using costs effective on February 14, 1989, and parity for concentrate pricing:

The cost of CFC Formula I is:

1.11 x 1.08 [(57% x $0.79) + (40% x $0.89)]
= 1.20 [$0.450 + $0.189]
= $0.967 for the CFCs in a 1.11 lb can, with loss.

The cost of CFC/HCFC Formula (not listed in Section 4) is:

1.11 x 1.08 [(53% x $0.79) + (26% x $2.40) + (18% x $1.05)]
= 1.20 [$0.419 + $0.624 + $0.189]
= $1.478 for the CFC/HCFCs in a 1.11 lb can, with loss.

The cost difference would then be:

Diff. = $1.478 - $0.967 = $0.511 per 1.11 lb average size can.

Notes:

- Analysis assumes the current can sizes can hold an extra 7 volume percent of product, since Formula V has a lower density than the present CFC formula (Formula I). Otherwise, a larger can or a smaller net weight will be required, adding to the cost per pound.

- The CFC content is immediately reduced from 97% to 53%; i.e., a 45% reduction. Ozone depletion is then reduced by about 43.6%.

Prices for Formulas III and IV are not calculated here because the use of methylene chloride may be difficult in light of its possible human toxicity, and because, according to industry sources, the relatively low volatility of 1,1,1-trichloroethane--inhibited will lead to an inferior product for many applications.

Summary

The cost of converting CFC Formula I to:

55% CFC/45% HCFC Formula IVa in 1989 will increase by about $0.511 per can;

100% HCFC Formula V in 1993/1994 will increase by about $0.204 per can; and

100% HCFC (+ chlorinateds) Formulas III or IV will change slightly in 1989.

The cost of a factory conversion from a TYPE 1A to a TYPE 2A product in 1989 is estimated to be $150,000 per moderate speed line. Note that the option of going to a contract filler is available. Several firms now use a contract filler, and a converted line could be also used for other products of TYPE 2A compositions.

Lubricants--For Electric/Electronic Equipment

Formula IX (see Section 4) will be taken as the standard, filled to 0.76 lb per can.

> 5% Specialty Lubricant Blend
> 65% CFC-113
> 30% CFC-12

The 1989 cost of the CFC components of this formula is as follows:

Cost = 0.76 x 1.08 [(65% x $1.40) + (30% x $0.89)]

   = 0.821 [$0.91 + $0.267]

   = $0.966 per can (with an 8% manufacturing loss of propellants; the 8% is an estimate of the propellant lost during filling).

In Section 4, formulas are presented for 1,1,1-trichloroethane and meter-spray options. If these are ruled out, we are left with a 1989 conversion to a CFC/HCFC formula, followed in 1993 or 1994 by conversion to a HCFC formula.

The proposed 1989 "partial conversion" formulation is the following:

> 5% Specialty Lubricant Blend
> 70% CFC-113
> 25% HCFC-22

The 1989 cost of the CFC/HCFC components is as follows:

Cost = 0.76 x 1.08 [(70% x $1.40) + (25% x $1.05)]

   = 0.821 [$0.980 + $0.263]

   = $1.020 per can (with an 8% manufacturing loss of propellants).

Conversion in 1989 from CFC Formula IX to the CFC/HCFC formula will cost:

$1.020/can - $0.966/can = $0.054/can

The reduction in CFC content is 26.3% and the reduction in ozone depletion is 24.2%.

A second stage of the reduction process could occur in 1993 or 1994, when HCFC-141b should become available. See Section 4, Formula XII:

5% Specialty Lubricant Blend
72% HCFC-141b
23% HCFC-22

Using 1993/1994 prices, the cost becomes the following:

Cost = 0.76 x 1.08 [(72% x $3.00) + (23% x $2.00)]
= 0.821 [$2.16 + $0.46]
= $2.15/can (with an 8% manufacturing loss of propellants).

The standard CFC (Formula IX) composition must be recalculated for 1993 and 1994 prices, from Table 12, to determine any differential.

Cost = 0.76 x 1.08 [(65% x $4.00) + (30% x $2.00)]
= 0.821 [$2.60 + $0.60]
= $2.63/can (with an 8% manufacturing loss of propellants).

Conversion in 1993 or 1994 from CFC (Formula IX) to HCFC (Formula XII) will save:

$2.63/can - $2.15/can = $0.48/can.

The 1993/1994 pricing structure for CFCs and HCFCs is very speculative. Cost increases or savings in conversion represent differentials and could be subject to considerable error.

Summary

The cost of converting from CFC Formula IX to:

the 73.7% CFC/26.3% HCFC Formula will result in an increase (based on 1989 prices) of about $0.054/can; and

the 100% HCFC Formula XII will result in a decrease (based on 1993/1994 prices) of about $0.48/can.

The 1,1,1-trichloroethane and meter-spray formulas are not considered here because they may have very limited applications. However, they are the least costly by far.

The cost of a factory conversion from TYPE 1A to a TYPE 1A (HCFC) formula in 1989 would be negligible.

Lubricants for Pharmaceutical Pill and Tablet Manufacture

The optimum immediate conversion possibility is for:

Formula XIII		**Formula XIV**
5% Concentrate		5% Concentrate
70% CFC-113 (Specially purified)	to	65% CFC-113 (Specially purified)
25% CFC-12		10% HCFC-142b
		20% HCFC-22
TYPE 1B		TYPE 2B

The average can size is 0.86 lb net weight. Costs are calculated as follows:

Formula XIII: Cost = 0.86 x 1.08 [(70% x $1.40) + (25% x $0.89)]

    = 0.929 [$0.980 + $0.223]

    = $1.118/can.

Formula XIV: Cost = 0.86 x 1.08 [(65% x $1.40) + (10% x $2.40) +

    (20% x $2.00)

    = 0.929 [$0.910 + $0.240 + $0.400]

    = $1.440/can.

The differential then becomes about $0.322/can. Reduction in CFCs would be 32%, and ozone depletion would be reduced by approximately 31%. The same can size could probably be used for equal weights of the alternative formula, since there is only about a 3% reduction in product volume. The manufacturing conversion cost would be negligible.

The preferred 1993/1994 alternative would be Formula XVI (see Section 4):

5.0% Concentrate

55.0% HCFC-141b

30.0% HCFC-124

10.0% HCFC-22

TYPE 2B

Costs are compared with Formula XIII (using 1993/1994 prices) as follows:

Formula XIII: Cost = 0.86 x 1.08 [70% x $4.00 + 25% x $3.40]

    = 0.929 [$2.80 + $0.85]

    = $3.39/can.

Formula XVI: Cost = 0.86 x 1.08 [55% x $3.00 + 30% x $3.70 +
                    10% x $2.00]
                  = 0.929 [$1.65 + $0.20]
                  = $2.75/can.

The price decrease in going to the HCFC formula then becomes $0.64/can. This could be further decreased if a charge were made for extraordinary purification of the HCFC-141b volatile liquid component. The same size can could probably be used, with the same formula weight. The manufacturing conversion cost should be negligible in going from TYPE 1B to TYPE 2B.

The conversion of CFC Formula XIII to:

   the 68.4% CFC/31.6% HCFC Formula XIV will result in an increase (based on 1989 prices) of about $0.322 per average-size can for this product; and

   the 100% HCFC Formula XVI will result in a decrease (based on 1993/1994 prices) of about $0.64 per average-size can for this product. This decrease could dwindle, however, if the HCFC-141b requires special purification.

Contingency formulas, such as Formula XV, are discussed in Section 4 but are not priced here.

The conversion from a TYPE 1B to a TYPE 2B formulation should not have a significant impact in the manufacturing area.

Solvent-Cleaners. Dusters. and Coatings for Electric/Electronic Equipment

The standard formulation is the following:

Formula XVII

75% CFC-113
25% CFC-12

TYPE 1A

The standard can size is 0.51 lb.

The cost is calculated as follows:

Cost = 0.51 x 1.08 [(75% x $1.40) + (25% x $0.89)]
     = 0.551 [$1.05 + $0.223]
     = $0.701/can.

The only immediately available conversion is to the following formulation:

Formula XIX

70% CFC-113
16% HCFC-142b
14% HCFC-22

TYPE 2A

and to closely related analogs, if a maximum amount of CFC-113 and CFC-12 is to be replaced.

The cost is calculated as follows:

$$Cost = 0.51 \times 1.08 \ [70\% \times \$1.40 + 16\% \times \$2.40 + 14\% \times \$1.05]$$
$$= 0.551 \ [\$0.98 + \$0.384 + \$0.147]$$
$$= \$0.833/can.$$

The present differential then becomes about $0.132/can. The reduction in CFCs would be 30%. The product density would decrease by about 11%, possibly requiring a larger can for the same fill weight, or else a reduction in fill weight. The manufacturing conversion cost would be negligible.

The ultimate conversion, to an entirely CFC-free formula, could take place around 1993 or 1994 when the "future alternatives" become commercially available. The preferred formulation is Formula XII (see Section 4), although others are given, according to the availability of HCFC-123 and the desire to use propane as a cost-cutting option. Following is a comparison of the 1993/1994 prices of Formulas XVII and XXI.

Formula XVII (1993/1994)	Formula XXI (1993/1994)
75% CFC-113	60% HCFC-141b
25% CFC-12	30% HCFC-124
	10% HCFC-22

As before, the 1993/1994 cost is calculated as follows:

$$Cost = 0.51 \times 1.08 \ [75\% \times \$4.00 + 25\% \times \$3.40]$$
$$= 0.551 \ [\$3.00 + \$0.85]$$
$$= \$2.12/can.$$

For Formula XXI the cost is calculated as follows:

$$Cost = 0.51 \times 1.08 \ [60\% \times \$3.00 + 30\% \times \$3.70 + 10\% \times \$2.00]$$
$$= 0.551 \ [\$1.80 + \$1.11 + \$0.20]$$
$$= \$1.714$$

The differential then becomes $0.41/can. The reduction in CFCs would be 100%, and the reduction in ozone depletion would be about 96%. The formula would be about 10 to 15% lower in density than the CFC counterpart, thus a slightly larger can, or slightly reduced-weight formula/package may be needed. Because the product is being converted from a TYPE 1A to another TYPE 1A product--assuming a three-component blend is prepared by the supplier and added to the can by the filler--, there would be a negligible effect on manufacturing conversion costs. It is not recommended that the slightly flammable HCFC-141b (B.P. = 90°F) be handled as a separate concentrate. Because of distillation effects in storage tanks, the filling of Formula XXI should be as follows:

- From a three-component blend in a bulk tank that is never allowed to sink below 35 volume percent full; or

- From a bulk tank of a two-component blend of HCFC-141b/124 in a ventilated, explosion-proof gas-house, followed by separate gassing of HCFC-22 into the cans.

The second option would force a conversion from TYPE 1A to TYPE 3A, which could cost up to about $400,000 to $1,200,000 per moderate-speed line. This is not recommended unless the filler already has such a TYPE 3 facility in place.

Summary

The 1989 price of converting Formula XVII to interim Formula XIX will result in an average increase of $0.132/can (CFC content will decrease by 30%).

The ultimate 1993/1994 price of converting to Formula XXI will result in a decrease of about $0.41 per can (CFC content will then be zero).

Metered Dose Oral and Nasal Inhalation Pharmaceutical Drug Products (MDIDs)

Formulations vary, but the "standard" formulation for the popular microcrystalline suspended (drug) solids type is as follows:

FORMULA MDID

2% Drug and Excipient Concentrate
24% CFC-11
24% CFC-114
50% CFC-12

The average package size is 0.033 lb.

The 1989 cost of the CFC ingredients is as follows:

Cost = 0.033 x 1.08 [24% x $0.79 + 24% x $1.23 + 50% x $0.89]
        0.0356 [$0.190 + $0.295 + $0.445]
        $0.0331/canister

As discussed in considerable detail in Section 4, the industry has firmly rejected all currently available non-CFC propellants; therefore, there is no immediately available alternative whose price can be compared with that above.

The best future alternative formula for these products may be Formula XXV (the modification with 10% HCFC-124 propellant), which is as follows:

Formula XXV

2% Drug and Excipient Concentrate*
13% HCFC-123**
75% HFC-134a
10% HCFC-124

*Changed from 1.5% for a better comparison with Formula
MDID, above.

**Based on this slurrying agent having a sufficiently
low toxicity for the application. (Otherwise, use
Formula XXVII, which has 4.5% CFC-113, but is similar
in cost).

The 1993/1994 price for the CFC ingredients of the CFC Formula MDID is
calculated as follows:

Cost = 0.033 x 1.08 [24% x $3.00 + 24% x $4.00 + 50% x $3.40]
     = 0.0356 [$0.72 + $0.96 + $1.70]
     = $0.120/canister

The 1993/1994 cost of the HCFC and HFC ingredients in Formula XXV future
replacement is calculated as follows:

Cost = 0.033 x 1.08 [13% x $4.00 + 75% x $4.40 + 10% x $3.70]
     = 0.0356 [$0.52 + $3.30 + $0.37]
     = $0.149/canister

The cost increase for the HCFC replacement is then $0.029 per canister.

For manufacturing purposes, Formula XXV is a TYPE 1A composition. The
HCFC replacement formula is also a TYPE 1A, and all three HCFCs are nonflammable. Thus, manufacturing conversion costs should be minimal.

However, for these products, the cost of research, increased toxicology
studies, dealing with the FDA (NDA-amendment), and additional quality control
methods development would be considerable. They are estimated to be
$2,500,000 per product, for each of the approximately 21 products now on the
U.S. market. When this is added to the manufacturing, marketing, and other
costs, the industry total for converting is estimated to be about $60,000,000,

unless a more practical cooperative approach is used to qualify families of similar products.

Summary

No immediate or short-term conversion appears likely, especially in view of the lengthy product development, toxicological testing, and regulatory approval timeframe.

The 1993/1994 conversion will add a $0.029 cost increment to the CFC formulas. It will allow a 100% reduction in CFC content and a 99% reduction in ozone depletion potential.

About $60,000,000 in conversion costs is anticipated, industry wide.

The dissolved-drug forms of MDIDs have not been discussed (except briefly in Section 3), but their costs are comparable to those of the microcrystalline suspension forms. They represent less than 10% of the business volume and are provided by only two marketers, one of whom already has a microcrystalline suspension product in the line. An additional $3,900,000 conversion cost is estimated for these two marketers.

The total MDID conversion cost is therefore estimated to be about $64,000,000.

Contraceptive Vaginal Foams (Human Uses)

The current products average 0.16 lb net weight and contain about 3.2% CFC-114 and 4.8% CFC-12. A reasonable conversion would be to the hydrocarbon propellant form, using 4.00% Blend A-46, and 4.00% added water.

No rapid conversion is foreseen because of the usual 3- to 5-year FDA NDA time. Price comparisons based on 1993/1994 costs are therefore required.

Cost of CFC Formula XXXI:

Cost = 0.16 x 1.35 (loss) x [3.2% x $4.00 + 4.8% x $3.40]
     = 0.216 [$0.128 + $0.154]
     = $0.061/can

Cost of the Hydrocarbon A-46 Blend formula (Formula XXXII):

Cost = 0.16 x 1.25 (loss) x [4.0% x $0.23 + 4.0% x $0.005*]
     = 0.200 [$0.0092 + $0.0002]
     = $0.0019/can

* Cost of deionized water; U.S.P.

The conversion from a CFC to a hydrocarbon propellant would save $0.059/can in 1993/1994 prices.

This option, with the attendant cost savings, will appeal to those marketers who have their own line(s) already able to safely fill hydrocarbon propellants, or who have their product filled by contract fillers. Others will have to spend from $400,000 to $1,200,000 per moderate speed line, depending on the relative degree of safety desired or affordable.

For those marketers who wish to continue filling nonflammable propellants, or who are required by the FDA (Drug Division) to do so, the immediate alternative is the following:

Formula XXXIV

Concentrate	92.0%
Added DX-Water	1.5%
HCFC-142b	3.9%
HCFC-22	2.6%

TYPE 2A

The cost of this formula (minus the concentrate) is as follows:

$$Cost = 0.16 \times 1.30 \ (loss) \ [1.5\% \times \$0.005 + 3.9\% \times \$3.25 + 2.6\% \times \$2.00]$$
$$= 0.208 \ [\$0.0001 + \$0.1268 + \$0.0520]$$
$$= \$0.0372/can$$

The differential, in 1993/1994 prices, between CFC Formula XXXI and HCFC Formula XXXIV is $0.024/can. Formula XXXIV is less expensive.

Summary

Because of the FDA (NDA) requirement, no immediate or short-term conversions are practical. Marketers will probably submit both hydrocarbon and HCFC formulas, and the FDA will decide which is safest for the user.

A conversion from CFC to hydrocarbon A-46 will save about $0.059/can in 1993/1994 prices.

A conversion of CFC to HCFC will save about $0.024/can in 1993/1994 prices.

Conversion to hydrocarbon A-46 will allow a 100% reduction in CFC content and a 100% reduction in ozone depletion from this product. Similarly, a conversion to HCFC will constitute a 100% reduction in CFC content and a 97% reduction in ozone depletion expected from the product.

Mercaptan (Thiol) Warning Devices

The standard formula can be immediately switched to an HCFC one, as follows:

Formula XXXV   Formula XXXVI

2.0% Ethyl Mercaptan   2.0% Ethyl Mercaptan*
98.0% CFC-12   to   39.2% HCFC-22
                                        58.8% HCFC-142b

TYPE 1B   TYPE 2B

*Changed from 2.21%. The 2.21% was designed to allow
the same weight of ethyl mercaptan per unit volume of
formula, i.e., the same size can. Thus, the amount of
stenching agent per standard size can would stay constant. The change makes no real cost difference, yet
allows both propellants to be calculated at the 98%
level.

The cost difference can be developed as follows:

Diff. = 0.95 x 1.08 [(39.2% x $1.05 + 58.8% x $2.40) - (98% x $0.890]
      = 1.026 [$.4116 + $1.4112 - $.8722]
      = $0.975/can.

Thus, the replacement formula, which totally eliminates the CFC content,
costs about $0.975/can more in 1989. Ozone depletion from this source is
reduced by about 97%.

Fillers must handle the HCFCs as a pre-blend or install very costly
facilities for handling HCFC-142b as a flammable gas.

Intruder Alarm Device Canisters

The standard formula can be compared with the immediately available
alternative, as follows:

Formula XXXVII		Formula IXL
100% CFC-12		60% HCFC-142b
	to	40% HCFC-22
TYPE 1		TYPE 2

The cost differential can be presented as follows:

Diff. = 0.47 x 1.08 [(60% x $2.40 + 40% x $1.05) - (100% x $0.89)]
  = 0.5076 [$1.44 + $0.42 - $0.89]
  = $0.492/can.

Thus, the HCFC replacement unit costs about $0.492/can more in 1989 and totally eliminates the CFC content. Ozone depletion from this source is reduced by 97%.

Fillers must handle the HCFCs as a nonflammable pre-blend, or else install very costly facilities for handling flammable HCFC-142b.

Flying Insect Sprays for Food-Handling Areas

The CFC-based product has been replaced with a hydrocarbon-propelled counterpart. Formulas can be compared as follows:

Formula XL		Formula XLI
15% Concentrate A		15% Concentrate B
42.5% CFC-11	to	55% Deionized Water
42.5% CFC-12		30% Hydrocarbon Blend BIP
TYPE 1B		TYPE 3A

Note: The cost of Concentrate A is very close to that of Concentrate B. They are considered the same.

The cost differential can be presented as follows:

Diff. = 0.76 x 1.08 [(42.5% x $0.79 + 42.5 x $0.89) - (55% x $0.005 +
        30% x $0.180]
Diff. = 0.821 [$0.336 + $0.378 - $0.003 - $0.054]
      = $0.539/can

Therefore, the marketer would save $0.539/can in factory cost by con-
verting to the hydrocarbon version. The CFC would be eliminated and ozone
reduction would be 0% of the CFC package.

A very costly HCFC alternate, Formula XLII, was described in Section 4
but would probably never be used.

Flying Insect Sprays for Aircraft

The compositions of the current CFC-based product and of a proposed
HCFC-based alternative are as follows:

Formula XLIII	Formula XLIV
2% Toxicant Blend	2% Toxicant Blend
49% CFC-11 and	58% 1,1,1-Trichloroethane-Inhibited (4)
49% CFC-12	40% HCFC-22
TYPE 1A	TYPE 1A

The cost differential can be presented as follows:

Diff. = 0.48 x 1.08 [(49% x $0.79 + 49% x $0.89) - (58% x $0.36 +
        40% x $1.05)]

Diff. = 0.5184 [$0.387 + $0.436 - $0.209 - $0.420]
      = $0.100/can

Thus, the marketer would save $.100/can by changing to the HCFC-based formula. CFC usage would be eliminated and the ozone depletion would be reduced by about 97%.

A far more costly alternative using HCFC-22/HCFC-142b could be used if the 1,1,1-trichloroethane were not allowed for some reason.

Flying Insect Sprays for Tobacco Barns

This product probably does not exist in standard aerosol forms. For the 9 lb-, 15 lb-, and 25 lb-(net) cylinders now in use, CFC-12 could be replaced by HCFC-22 without fear of over-pressurization.

Considering that 40% HCFC will do the dispersant work of 50% CFC-12, a cost comparison can be made as follows:

Cost of 0.40 lb of HCFC-22 is 1.05 x $1.05 x 0.40 = $0.441
Cost of 0.50 lb of CFC-12 is 1.05 x $0.89 x 0.50 = $0.467

Differential: $0.026

This means that the cost of propellant gas per pound of insecticide is only $0.026 more for the alternative HCFC-22 formula.

Another approach is to consider the propellant cost in terms of each pound of Vapona dispersed. In this case, the cost of the HCFC-22 is less, since $0.735 of HCFC-22 sprays one pound of Vapona; whereas, $1.168 of CFC-12 is required to perform the same task.

Aircraft Maintenance and Operation Sprays

A number of different products fall into this (presently CFC) category. The lubricants, dusters, and insect sprays have already been reviewed above.

In Formula XLVI, 25% HCFC-22 replaces 20% CFC-12 in a hydro-alcoholic disinfectant/deodorant system for the control of mold and mildew. For a 14 Av.oz. average can the cost increase for the alternative formula calculates as follows:

Diff. = 0.89 x 1.08 [25% x $1.05 - 20% x $0.89]
     = 0.959 [$0.2625 - $0.178]
     = $0.081/can

A much higher-priced, essentially anhydrous HCFC-141b/HCFC-124 formulation is Formula XLVII, but its efficacy would probably vary with the available humidity in the air and, generally, Formula XLVI would be more effective.

Military Aerosols

The following analysis of Formula XLIII versus Formula XVIV applies to military flying insect sprays, as do the analyses of lubricants, dusters, etc.

The largest military purchases are for the 10-Av.oz. 2% toxicant, 98% CFC-12/11 insecticides. These can be compared with HCFC versions as follows:

Formula XLVIII		Formula XLII
2% Toxicants		2% Toxicants
49% CFC-11	to	42% HCFC-22
49% CFC-12		56% HCFC-142b
TYPE 1A		TYPE 2A

The 1989 cost differential can be determined as follows:

Diff. = 0.64 x 1.08 [(49% x $0.79 + 49% x $0.89) - (42% x $1.05 +
      56% x $2.40)]

$$= 0.691 \ [\$0.387 + \$0.436 - \$0.441 - \$1.34]$$
$$= -\$0.662/can$$

Thus, a conversion to Formula XLII would cost $0.662 more per average-size can at the factory cost level. Formula XLII contains no CFCs and its ozone depletion potential would be 3% that of Formula XLVIII.

CFC-115 for Aeration of Food Products--Certain Limitations Applied

This market no longer exists, according to DuPont, the sole CFC-115 supplier. No future revival is anticipated.

EXCLUDED PRODUCTS

Tire Inflators

The prevalent formula types can be compared as follows:

<table>
<tr><td>Formula L</td><td></td><td>Formula LI(a)</td></tr>
<tr><td>45% Water-based Sealant</td><td></td><td>45% Water-based Sealant</td></tr>
<tr><td>22% CFC-114</td><td>to</td><td>30% Additional Water</td></tr>
<tr><td>33% CFC-12</td><td></td><td>25% Hydrocarbons A-46</td></tr>
<tr><td>TYPE 1A</td><td></td><td>TYPE 3A</td></tr>
</table>

Formula LI

45% Water-based Sealant
20% Additional Water
23% HCFC-142b
12% HCFC-22

TYPE 2A

Cost developments are as follows:

Formula L: Cost = 0.93 x 1.08 [22% x $1.23 + 33% x $0.89]
             Cost = 1.004 [$0.271 + $0.205]
             Cost = $0.478/can (for CFCs)

Formula LI(a):   Cost = 0.93 x 1.08 [30% x $0.005 + 25% x $0.18]
                 Cost = 1.004 [$0.0015 + $0.0450]
                 Cost = $0.047/can (for extra water and A-46)

Formula LI: Cost = 0.93 x 1.08 [20% x $0.005 + 23% x $2.40 +
                  12% x $1.05]
              Cost = 1.004 [$0.001 + $0.552 + $0.126]
              Cost = $0.682/can (for extra water and HCFCs)

Differentials then become as follows:

Formula L to LI(a): a decrease of $0.431 per can (factory cost).

Formula L to LI: an increase of $0.204 per can (factory cost).

Diamond Grit Spray

Almost no substantive information on this product has been found. It is
an exotic abrasive spray used for high-technology applications. Approximately
250,000 units a year are used.

Polyurethane Foams

Elimination of the present 32% CFC-12 formulation is suggested in favor
of a 32% HCFC-142b/22 (17:15) blend.

Cost comparison is as follows:

Cost = 1.01 x 1.08 [(32% x $0.89) - (17% x $2.40 + 15% x $1.05)]

= 1.09 [$0.285 - $0.408 - $0.158]

= -$0.306/can (factory cost) differential

This will eliminate CFCs from this source and reduce ozone depletion by this product by 97%.

Chewing Gum Removers

The suggested transition is from a 100% CFC-12 formulation to a blend of 44% HCFC-22 and 56% HCFC-142b (see Formula LIII).

The cost comparison is as follows:

Diff. = 0.59 x 1.08 [100% x $0.89 - (44% x $1.05 + 56% x $2.40)]

Diff. = 0.6372 [$0.89 - $0.462 - $1.34] = $0.581/can = factory cost.

The switch to HCFCs will cost $0.581 more per average can at the factory and eliminate CFC emissions from this source. A 97% reduction in ozone depletion from these products will result. (As of July 1, 1989, DuPont and Allied Signal will no longer sell CFCs for this application.)

Drain Openers

This product is no longer on the market and will probably not be revived.

Chillers - Medical

The present and comparable future HCFC formulas can be compared as follows:

Formula LIV(a)   Formula LV

10% Ethanol - Anhydrous   10% Ethanol - Anhydrous
50% CFC-114   to   50% HCFC-142b
40% CFC-12   40% HCFC-22

              TYPE 1B   TYPE 2B
Formula IV(a): Cost = 0.39 x 1.08 [50% x $1.40 + 40% x $0.89]
                     = 0.4212 [$0.70 + $0.356]
                     = $0.445/can (for the CFCs) = factory cost

Formula IV: Cost = 0.39 x 1.08 [50% x $2.40 + 40% x $1.05]
                  = 0.4212 [$1.20 + $0.42]
                  = $0.6823/can (for the HCFCs) = factory cost
The differential in price is $0.237/can.

A transition to the HCFC formula would eliminate CFCs in this product
and reduce ozone depletion from this source by 97%.

Non-Electric/Electronic Dusters

A transition from 100% CFC-12 to a blend of 60% HCFC-142b and 40% HCFC-
22 is suggested, as follows:

The cost differential should be as follows:

Diff. = 0.43 x 1.08 [(60% x $2.40 + 40% x $1.05) - (100% x $0.89)]
Diff. = 0.4644 [$1.440 + $0.42 - $0.890]
Diff. = $0.450/can = factory cost

Microscope Slide Cleaners and Related Products

The following transition is suggested:

Formula LV(a)		Formula LVI
75% CFC-113		72% 1,1,1-Trichloroethane - Inhibited (4)
25% CFC-12	to	28% HCFC-22
TYPE 1A		TYPE 1A

Cost comparisons are as follows:

Diff. = 0.39 x 1.08 [75% x $1.40 + 25% x $0.89 - (72% x $0.36 +
        28% x $1.05]

Diff. = 0.4212 [$1.050 + $0.223 - $0.242 - $0.294]

Diff. = $0.310/can

This change would eliminate CFCs from this source and limit ozone depletion by these products to about 3% the current level.

No special manufacturing costs would be incurred. The HCFC formula would be about 3 to 4% lower in density, so changes in can size or net weight would probably not be required.

Solvents - Medical

The current formula cannot be replaced with any combinations of HCFCs or other formulations now available. In 1993/1994 a replacement formula could be provided. The two are compared, using 1993/1994 pricing, as follows:

Present Formula	Formula LIX
3% Concentrate	3% Concentrate
72% CFC-113 (Purified)	41% HCFC-141b
25% CFC-12	40% HCFC-123
	16% HCFC-22
TYPE 1A	TYPE 2A

Present Formula: Cost = 0.51 x 1.08 [72% x $4.00 + 25% x $3.40]

= 0.551 [$2.880 + $0.850]

= $2.055/can (factory cost) for CFCs

Formula LIX: Cost = 0.51 x 1.08 [41% x $3.00 + 40% x $4.00 +

16% x $2.00]

= 0.551 [$1.23 + $1.60 + $0.32]

= $1.736/can (factory cost)

The differential is thus $0.319/can. (The HCFC formula is less costly.)

A temporary reformulation can be created by replacing the CFC-12 with 18% HCFC-22.

The calculation is as follows:

Cost = 0.51 x 1.08 [82% x $1.40 + 18% x $1.05]

= 0.551 [$1.148 + $0.189]

= $0.7367/can

The differential then becomes $0.7367/can (CFC/HCFC) - $0.6770/can (CFC) = $0.0597/can.

Boat Horns

The present CFC-12 product costs are as follows:

$$Cost = 0.77 \times 108 \ [100\% \times \$0.89]$$
$$= 0.8316 \times \$0.89$$
$$= \$0.7401/can$$

A conversion to a 60% HCFC-142b + 40% HCFC-22 formula can be suggested, and the cost would then be:

$$Cost = 0.77 \times 1.08 \ [40\% \times \$1.05 + 60\% \times \$2.40]$$
$$= 0.8316 \ [\$0.420 + \$1.440]$$
$$= \$1.547/can$$

The differential between the two formulas is thus $0.807/can.

Contingency Products

This category includes miscellaneous uses too small to be recognized by CFC suppliers and the industry at large. It also includes unauthorized or illegal applications of CFC propellants to various aerosol products.

It doesn't include CFCs as a part of Halon fire extinguisher formulations or CFC aerosol-size refrigeration/air conditioner refill units, although these are surveyed as a part of the aerosol product volume per year; and it does not include the aerosol-size sterilants, based on about 10% ethylene oxide "tamed" in 90% CFC-12.

These products have a sales unit volume of about 410,000 cans. The can fill is estimated to consist of 0.50 lb CFC-based ingredients. The category is handled as if the products were medical solvents.

Cost — Present CFC cost: $0.6770/can
    — Proposed 82% CFC-113 and 18% HCFC-22 formula cost: $0.7367/can

Cost differential: $0.0597/can.

Summary

Table 14 summarizes the short-term (1989) and long-term (1993/1994) price differentials of the CFC-based products described in detail in this section.

The double-dashes (--) in Table 14 signify that short-term, fully corrective measures can be taken, and that the longer-term availability of future "alternative" propellants is of no consequence. Triple dashes (---) signify that there is no definable short-term reformulation activity.

Where reasonable evidence exists that a product has been discontinued, zeros (0) in the unit volume column have been used to signify that there is no known or suspected production of the products. For the tire inflator, the cost of only one formula option has been calculated on a yearly basis: the one with a blended hydrocarbon (A-46) propellant. The alternative (nonflammable) version is too costly to consider.

In terms of manufacturing changeovers, such as the purchase of new tanks, new gas-houses, new monitoring equipment, employee education, etc., in-house fillers would spend an estimated total of $15,000,000 for non-pharmaceutical items.

Pharmaceutical and medical products are more difficult to revise, requiring more testing, more development, linkages with the FDA (NDA), in addition to manufacturing revisions. Total costs are estimated to be approximately $70,000,000, unevenly divided between marketers who contract-fill and those who self-fill. Twenty-two or twenty-three products are involved in the MDID area, for an increment of about $64,000,000. The rest are involved in the contraceptive foam area, for an increment of about $6,000,000.

TABLE 14.   DIFFERENTIAL COSTS FOR SHORT-TERM AND LONG-TERM CONVERSIONS

Product Category	Unit Volume	Present CFC Cost Per Can ($)	Differential Per Can (Short-Term)($)	Differential Total (Short-Term)($)	Differential Per Can (Long-Term)($)
Mold Releases	1,320,000	0.967	0.511	674,500/yr	0.204 increase
Lubricant E/E	2,320,000	0.966	0.054	125,300	0.480 decrease
Lubricant Tab.	1,930,000	1.118	0.322	621,500	0.640 decrease
Solvent Cleaners, Dusters for E/E Equipment	10,900,000	0.701	0.132	1,438,800	0.410 decrease
MDID Inhalants	107,000,000	0.0331	---a	---	0.029 increase
Contraceptives	6,940,000	0.0181	---a	---	0.059 decrease b
Mercaptan Warn.	90,000	0.895	0.975	88,000	--
Intruder Alarm	1,210,000	0.452	0.492	595,300	--
Pesticide-Food	120,000	0.586	0.539	64,700	--
Pesticide-Aircraft	580,000	0.426	(0.100)	(58,000)	--
Pesticide-Tobacco Barns	60,000	0.467/lb	(0.026)/lb	(1,560)	--
Aircraft Maintenance	110,000 500,000	0.171 0.966	0.081 D/Dc 0.054 Lubr.	8,100 27,000	-- --
Military Aerosols	1,960,000	0.568	0.662	1,298,000	--

(Continued)

TABLE 14. (Continued)

Product Category	Unit Volume	Present CFC Cost Per Can ($)	Differential Per Can (Short-Term)($)	Differential Total (Short-Term)($)	Differential Per Can (Long-Term)($)
Tire Inflators	1,650,000	0.478	(0.431) Hydr. 0.204 HCFC	(711,200) Hydr.	--
CFC-115 (Food)	0	-	--	--	--
Diamond Grit	250,000	-	--	--	--
Polyurethane Foam	9,170,000	0.310	0.306	2,806,000	--
Chewing Gum Remover	470,000	0.567	0.581	273,100	--
Drain Openers	0	-	--	--	--
Chillers - Medical	710,000	0.445	0.237	168,300	--
Non-E/E Dusters	860,000	0.413	0.450	387,000	--
Boat Horns, etc.	2,280,000	0.740	0.806	1,837,700	--
Microscope Slides	120,000	0.536	0.310	37,200	--
Solvents - Medical	1,090,000	0.677	0.060	650,700	0.319 decrease
Contingencies	410,000	0.677	0.060	246,000	--

[a] No 1989 conversion data can be provided for the Metered-Dose Inhalant Drug (MDID) or Human Contraceptive products, since any change requires FDA (NDA) approval, which may take 3 to 5 years from the time of application and submittal of the research data. The FDA may or may not approve, because of the theoretical flammability.

[b] This is based on using the recommended hydrocarbon propellant blend version.

[c] This is the mold/mildew disinfectant/deodorant; one of several products used by aircraft maintainers/operators. All the others have been grouped together under the heading of the very large lubricant area.

PROCEDURES FOR CHANGING FROM CFC TO ALTERNATIVE FORMULATIONS

The progression of events for changing formulas has been covered to some extent earlier in this and other sections, but it is more fully defined in the following paragraphs.

Research Phase

- Order new propellants, prepare and test samples.

- Develop analytical methods for quality assurance.

- Develop specifications for quality control.

- Develop can/valve compatibility with product, including weight loss data, temperature sensitivity, reductions in active ingredient assay, and any other changes.

- Provide Material Safety Data Sheets on chemicals, concentrate, and complete formula.

Manufacturing Preparations

- Test new product(s) in pilot facilities, if available.

- Order new equipment as necessary.

- Install and perform "shake-down" on new equipment.

- Obtain spare parts, manuals, and educate employees on the use of new equipment.

- Install explosion-resistant, highly-protected gas house if dealing with TYPE 2 or TYPE 3 formulations.

Sales and Marketing

- Prepare new cost information:

  -- Factory cost data;
  -- Selling cost data;
  -- Manufacturing capitalization cost data;
  -- Product re-development cost data; and
  -- Regulatory liaison cost data.

- Prepare new product labels, literature, advertising, and time-
  tables.

- Educate sales personnel as to advantages/disadvantages, costs, etc.

- Educate distributors.

- Prepare for orderly transition in the marketplace.

Production

- The complexity of reformulating will vary with the product,
  increasing with pharmaceutical drug types. The list of procedures
  is not meant to be complete, merely indicative.

SUMMARY

For a group of 26 product categories, the costs of short-term and/or
long-term formula revisions have been calculated for current and optimum
formulas, and later extended into the product cost increases or decreases
shown in Table 14.

Data on product size, restrictions, estimates, etc. have been included,
as well as predictions about the reduction of CFC content and the corresponding reduction of stratospheric ozone depletion by CFC source.

Complete elimination of CFCs is considered possible on a short-term basis for about 18 of the 26 product categories covered.

Data on the cost of manufacturing/research/quality control/regulatory activities have been presented. These amount to a total of about $15,000,000 for non-drug products and about $70,000,000 for drug products. Elements that could affect these numbers include the following:

- Marketer/fillers leaving the marketplace;

- Marketer/fillers going entirely to contract fillers;

- Possible cost savings from cooperation in the FDA/NDA process;

- Coalitions allowing several marketers to sell one FDA-approved formula/package; and

- Reduction in sales volume, so that smaller lines can be used.

Considerable formulation work and related activities are being conducted at this time, and the industry foresees a 25% reduction in the availability of CFCs by July 1, 1989, caused by implementation of the first phase of the Montreal Protocol.

6. Conclusions

Sections 4 and 5 have examined reformulation possibilities for 26 categories of CFC aerosol products that have enjoyed exempted or excluded status relative to the CFC bans and limitations placed on the aerosol industry by the U.S. EPA and FDA in 1978. According to the particulars of their use, some of these products may be easily reformulated to exclude CFCs, and some may be difficult to reformulate. Several cannot be reformulated with existing alternative propellants, but must await the commercialization of the four "future alternative" propellants described in Sections 4 and 5. Others must not only wait for the development of these new propellants, but must then go through the 3- to 5-year NDA procedure imposed by the FDA.

AEROSOL USES FOR WHICH CFCS ARE DIFFICULT TO ELIMINATE (AND POSSIBLE INTERIM REFORMULATIONS)

The seven categories of aerosol uses from which CFCs are most difficult to eliminate are discussed in this section. However, partial or interim reformulations of some categories to decrease CFC use are also noted. Table 15 lists the categories and the U.S. consumption of CFCs from these categories in millions of pounds per year. The perceived need for CFCs in these products is based on the lack of available alternatives that can completely replace the CFCs at the present time; however, approximately 40 percent of the CFCs now used in exempted or excluded U.S. aerosol products can be immediately replaced. Further, alternative non-CFC formulations for the seven remaining CFC-dependent categories are judged technically feasible, pending the commercial availability of four "future" HCFCs and HFCs.

117

TABLE 15.   AEROSOL USES FOR WHICH CFCs ARE DIFFICULT TO ELIMINATE

CFC Aerosol Product	U.S. Consumption of CFCs (MM lbs/yr)
Certain Mold Releases	1.5
Lubricants--for Electric/Electronic Applications	1.9
Lubricants--for Pharmaceutical Pill and Tablet Presses	1.0
Solvent cleaners, Dusters, etc. for Electronic/Electric Equipment	6.0
Metered-Dose Inhalant Drugs (MDID)	3.9
Contraceptives for Human Use	0.1
Solvents--Medical	0.6
	———
TOTAL	15.0ᵃ

ᵃThis is (15.0/25.5) = 58.8% of the present usage level.

Mold Release Agents

Key necessary attributes are product purity, surface-spray characteristics, fast evaporation of all but the lubricant, and nonflammability.

The nonflammability attribute relates to safety during large-scale intermittent use in unprotected areas. Using additional product (as silicones) promotes surface-coating characteristics but adds more lubricant than needed, causing voiding problems in the molded pieces. The use of volatile solvents promotes surface-coating characteristics, but if they are flammable the total product will become flammable. The nonflammable solvents are restricted to methylene chloride and, possibly, 1,1,1-trichloroethane (4). The latter is not quite volatile enough for most uses, and traces contaminate molded plastics and elastomers because of its solvency. Methylene chloride is effective but is a possible mutagen. This solvent can remain dissolved to some degree in the deposited silicone film, and may then craze or matte the surface of molded parts. It is an extremely strong solvent. See formulations III and IV in Section 4.

The idea of using a meter-spray valve has been discussed with one mold release marketer; however, employees operating molding machines tend to grossly overuse these more costly products. Also, meter-spray valves do not work well with strong solvents.

One reformulation approach is to revise the formula to a combination CFC/HCFC type, thus reducing the CFC usage. For example, by using:

	5% Silicone		5% Silicone
NOT:	60% CFC-11	BUT:	3% Isopar C (Heptanes)
	35% CFC-12		22% HCFC-22
			70% CFC-11

CFC consumption could be decreased 26% (or 0.39 MM lbs per year).

CFC-11 will be a necessary ingredient of mold releasers until it can be replaced with a suitable nonflammable, very volatile liquid such as HCFC-123; B.P. = 82°F.

The production of molded plastic or elastomeric parts affects many industries, and certainly the electric and electronics industry. A machine, for example, may produce parts for a "sensitive" industry one day and a "nonsensitive" one the next. Providing different lubricants for each of the two would be hard or impossible. Electric and electronic parts must be perfectly molded and carry essentially no embedded or impregnated impurities.

Lubricants for Electric/Electronic Uses

Key necessary attributes are purity, surface-spray characteristics, intrinsic nonflammability and reasonably fast evaporation of all but the lubricant. CFC-113 is a preferred carrying agent. Unlike other chlorinated solvents, such as methylene chloride and 1,1,1-trichloroethane (4), it is not readily pyrolized to form corrosive agents that could change resistances and otherwise harm delicate equipment.

Meter-squirt applications might allow the inclusion of a higher percentage (to 70%) of lubricant per aerosol can, which may be a viable solution (with HCFC-22 propellant) for those applications that can be handled by means of an extension tube that fits into the aerosol valve button. Research needs to be conducted in this area. See the discussion of Formulas XI and X in Section 4.

As shown below, to reduce CFC consumption the CFC-12 can be replaced with an appropriate amount of HCFC-22:

5% Specialty Lubricant		5% Specialty Lubricant
65% CFC-113	TO:	73% CFC-113
35% CFC-12		22% HCFC-22

This will reduce the CFC content by 23% and consumption by 0.44 MM lbs per year. It may be possible to restrict the CFC product to electronic uses, and use non-CFC sprays for ordinary electric motors, crude relays, and other items that do not require a high degree of purity.

The ultimate elimination of CFC-113 from lubricants requiring it depends on the future availability of liquid HCFCs, such as HCFC-123 (nonflammable) and HCFC-141b (slightly flammable). The latter would have to be used with a sufficient amount of HCFC-22 to eliminate any intrinsic flammability of the complete product.

Another possible interim formulation is 5% Lubricant, 55% CFC-113, and 40% HCFC-142b. If this formula tests out successfully, the CFC content would be reduced by 42.1%, or 0.80 MM lbs/yr.

Lubricants for Pharmaceutical Pill and Tablet Manufacture

This product is similar to the previous one. It normally consists of 70% CFC-113 and 15% CFC-12. (See Formula XIII in Section 4.)

Key attributes are high-purity, nonflammability and surface-coating action; everything but the lubricant evaporates quickly from the dies.

A short-term, partial conversion can be made using the following formula:

Formula XIV (see Section 4)

5% Concentrate (FDA Approved)
65% CFC-113
10% HCFC-142b
20% HCFC-22

This change would effect a 32% reduction in CFC content, for a reduction in CFCs used of 0.32 MM lbs per year.

Complete conversion, which will require the use of future alternative HCFC-123 (nonflammable) and/or HCFC-141b (slightly flammable), is several years away.

No presently available nonflammable solvents can begin to compare with CFC-113 for this application. The use of others would cause die contamination and inclusion of extraneous materials in the tablets and pills. The high-efficiency transfer of lubricant to the press dies would be impaired as well.

Solvent-Cleaners, Dusters and Coatings for Electric/Electronic Equipment

Electronic circuitry chillers and related products should also be included in this extremely large category, which accounts for 23% of all CFC aerosol materials.

Again, CFC-113, while unique in its selective solvent action on greases and oils, does not harm plastic or elastomeric components of circuit boards and other sensitive equipment. It also has the appropriate levels of volatility and purity, and it is nonflammable.

Instead of the usual formulation, 75% CFC-113 and 25% CFC-12 (Formula XVII, Section 4), a short-term partial conversion to Formula XIX could be made:

70% CFC-113

16% HCFC-142b

14% HCFC-22

In this case, the CFC reduction is 30%, for a decrease in CFCs consumed of 1.8 MM lbs per year.

The final conversion will have to await the availability of HCFC-123 or of HCFC-141b. For strategic planning purposes, tests should be undertaken with samples of these HCFCs to see if they can replace CFC-113.

The chiller sub-category uses straight CFC-12, possibly amounting to 20% of the CFCs used in the main category. The utility of the nonflammable blend, HCFC-22/142b (40:60), should be tested to discover any problems, such as the minor flammability of a surface from which much of the HCFC-22 (nonflammable) has evaporated. No other options are currently available. A future option may be 20% HCFC-124 and 80% HFC-134a.

Metered-Dose Oral and Nasal Inhalation Pharmaceutical Products

This industry can make no progress toward reformulation with the presently available alternative propellants. Studies with laboratory samples of the potential new propellants should be encouraged, assuming the toxicological tests now being conducted ultimately permit the marketing of these propellants.

As soon as the toxicology tests are successfully completed, the industry can approach the FDA and begin the NDA process, which will take 3 to 5 years to complete. One problem is that several firms have NDAs that are 20 years old, and they no longer have a technical staff able to develop a new product that would meet all the requirements of a revised NDA.

This is a large industry, producing 107 million units of inhalants a year. Marketers claim their product saves thousands of lives a year and makes other lives more bearable. If the industry works now on future alternatives, instead of waiting until 1993 when the toxicological tests will probably be completed, the length of time needed for completing the NDA amendment process and going to market with a new product can be significantly shortened. The use of discontinued FC-C318 (DuPont) should also be considered.

Contraceptive Vaginal Foams (Humans)

Reformulation with hydrocarbon alternatives could be completed very quickly if permitted by the FDA, at which point the FDA NDA amendment process could commence.

It is difficult to assess the "essentiality" of foam insert contraceptives (spermicides). It may be a highly discretionary alternative to oral contraceptives or other methods, or it may be a physical or emotional necessity as viewed by some users.

In any event, the quantity of CFCs consumed in this product category is very low--below 0.1 MM lbs per year.

Solvents - Medical

These products are generally for gauge bandage adhesives and adhesive removers. The primary silicone-adhesive supplier (Dow-Corning) has stated that, for technical reasons, no solvent besides CFC-113 is acceptable. The reason(s) is unknown. They may not have considered the future CFC alternatives. It may be that a complete conversion will have to await the commercialization of HCFC-141b (with a similar solvency profile).

In the short term, since present formulas are propelled with either 25% CFC-12 or 5% CO_2 gas, it may be practical to gain a partial conversion by using 18% HCFC-22 as the propellant. A further reduction in CFC consumption may be made by using 11% CFC-142b with 14% HCFC-22.

In the latter case, formulas with 90 to 95% CFCs will be replaced with one containing 70% CFC-113. This will reduce the CFC content by about (22.5/92.5) = 24%, decreasing the quantity of CFCs required for this product.

POTENTIAL FOR REDUCTION OF CFC USE IN EXEMPTED AND EXCLUDED AEROSOLS

The CFC usage data presented in Table 15 can now be augmented to include CFC reductions that can be made in the short term by partially reformulating certain products. Table 16 shows possible reductions projected in CFC consumption in the U.S. These reductions are based on the following scenarios:

TABLE 16.   PROJECTED ANNUAL CFC CONSUMPTION IN THE U.S.
UNDER DIFFERENT SCENARIOS (MM LBS/YR)

CFC Products	Present Consumption	1990 Usage Scenario One	1990 Usage Scenario Two	1995 Usage Scenario Three	2000 Usage Scenario Four
Mold Releases	1.50	1.50	1.11	0.00	0.00
Lubricants-E/E	1.90	1.90	1.10	0.00	0.00
Lubricants-Tablets	1.00	1.00	0.68	0.00	0.00
Solvents-E/E	6.00	6.00	4.20	0.00	0.00
MDID Inhalants	3.90	4.00	4.00	5.25	0.50
Contraceptives	0.10	0.10	0.10	0.00	0.00
Solvents-Medical	0.60	0.60	0.46	0.00	0.00
ALL OTHERS	10.50	0.00	0.00	0.00	0.00
	25.50	15.11	11.65	5.25	0.50

Scenario One:

The use of CFC-11, CFC-12, CFC-113, CFC-114, and CFC-115 would no longer be permitted for aerosol products currently exempted or excluded from previous regulations restricting their use, with the following exceptions:

- Release agents for molds used in the production of plastic and elastomeric materials;

- Non-consumer articles used as cleaner-solvents, lubricants, or coatings for electrical or electronic equipment, including freezants;

- Lubricants for rotary tablet press-punch machines;

- Solvents for medical purposes; and

- Metered-dose inhalant drugs, contraceptive vaginal foams and other human drugs as authorized by the FDA.

Scenario Two:

In addition to the items in Scenario One, for the first four products listed above, the use of CFC-11, CFC-12, CFC-114, and CFC-115 will no longer be permitted, and the amount of CFC-113 in such products will be limited to 75% by weight of the total formulation.

Scenario Three:

All the above items are included in this scenario, and the "future alternative propellants" must be capable of replacing CFC-113 and CFC-11. Also, the FDA must approve hydrocarbon use in contraceptives.

Scenario Four:

Finally, in this scenario all the above items apply, as well as the following: 1) there must not be any unusual delays in FDA's NDA Amendment process, and 2) the respective industries must proceed with the necessary research without delay.

Table 17 compares the reductions in CFCs called for by the Montreal Protocol with the reductions shown in Table 16.

NEAR-TERM CFC REDUCTIONS

The nonflammable blend of HCFC-22/142b (40:60) has a pressure of about 63 psig at 70°F, air free, and thus compares with CFC-12, which has a pressure of 70.69 psig at 70°F, air free. Since the blend is not quite as high in pressure, yet has an average molecular weight (and thus dispersive volume) very close to that of CFC-12, it may be necessary to increase the volume by 5 to 10% above the amount of CFC-12 when replacing it. This is illustrated by the following reformulation:

MOLD RELEASE AGENTS

CFC Formula	HCFC/CFC Formula
3% Concentrate	3% Concentrate
57% CFC-11	54% CFC-11
40% CFC-12	28% HCFC-142b
	15% HCFC-22

This shows how conveniently the CFC-12 can be replaced with an existing propellant blend. Formulation chemists have additional latitude they can explore to develop certain advantages; e.g., see the following formulas:

TABLE 17.   COMPARISON OF MONTREAL PROTOCOL AND QUICKEST
REASONABLE CFC REDUCTION

Protocol	Amount of CFC Propellants Used for U.S. Aerosols			
	Early 1989	Late 1989	1994/1995	1999/2000
Montreal Protocol	25.5 MM lbs	20.4[a]	16.3[a]	10.2[a]
Scenarios 1-4 (see Table 16)	25.5 MM lbs	15.0 or 11.7[b]	5.25	0.0

[a]Based on the levels of CFCs consumed in 1986 by the U.S. aerosol industry.
Actual reductions would consider total U.S. CFC consumption for all
applications.

[b]To achieve a reduction in CFCs to 15.0 MM lbs, the CFCs in all nonessential
aerosols must be replaced with HCFC, HFC, or hydrocarbon alternatives.

To achieve a reduction to 11.7 MM lbs, in addition to the above substitutions, as much HCFC or other propellants would have to be added to non-drug,
"essential" products as possible, replacing all the CFC-12 content and
perhaps minor amounts of CFC-113 or other CFCs. (This would be difficult to
accomplish in less than 1 to 1-1/2 years.)

The cost of all reformulated products would increase, in many cases quite
drastically, as shown in Section 5.

<u>Formula A</u>	<u>Formula B</u>
3% Concentrate	3% Concentrate
40% CFC-11	60% CFC-11
57% HCFC-142b	21% HCFC-142b
	16% HCFC-22
Compared with the HCFC/CFC	Compared with the HCFC/CFC
formula, increases cost, lightens	formula, reduces cost,
density, minimizes CFC-11.	increases density slightly,
	and increases CFC-11.

Extending the rationale behind Formula B would result in the following formulas:

<u>Formula C</u>	<u>Formula E</u>	<u>Formula D</u>
3% Concentrate	3% Concentrate	3% Concentrate
75% CFC-11	91% CFC-11	88% CFC-11
22% HCFC-22	6% CO_2	9% Propane A-108
Less costly	Still less costly	<u>Least</u> costly

The reduction of CFCs would not be as significant in these formulations.

A final approach, which may not be practical for every application, would use CFC-113 in place of the CFC-11, as shown below:

<u>Formula F</u>

3% Concentrate
35% CFC-113 (Purified)
62% HCFC-142b

Two reasons for using this approach are as follows:

* A heavy, surface-coating spray can be made using less CFC-113 than CFC-11; and

* CFC-113 is reported to be approximately 80% as damaging to the stratospheric ozone layer as CFC-11.

The CFC-113 is sufficient to quell any minor flammability of the CFC-142b during product use, making it nonflammable. Also, compared with the CFC Formula, the CFC content is reduced by (62/97) = 64%, and the ozone depletion potential is reduced by:

$$1 - \frac{0.80 \times 35\%}{1.00 \times 97\%} = 71.1\%$$

This is, however, an unusually costly formula, about \$1.22/lb of CFC/HCFC higher than the CFC formula.

This exercise shows that CFC-12 can be replaced with the nonflammable blend of HCFC-22/142b (40:60) or with some other blend of these alternative propellants, depending on concentrate and use factors. It is also desirable to replace the CFC-11 with 10-20% less CFC-113 and gain the added ozone protection. The drug aerosols (inhalants and contraceptives) are overseen by the FDA (NDA) and cannot be reformulated in this fashion.

Studies that should be conducted before further restrictions on CFC aerosols are announced include the following:

1. Offering formulation/packaging advice to affected marketers and seeking their feedback on product performance, manufacture, and standards compliance. For example:

    a. Will a proposed product sound a boat horn loudly enough, long enough, and without icing up or other complications?

b. Can a proposed product be directly gassed with HCFC-22 in essentially all pressure fillers (gassers) without blow-by, vapor-locks, gasket deterioration effects, etc.?

c. Will the formulations be considered satisfactory while limited by the Department of Transportation (DOT) to pressures of 180 psig at 130°F?

d. If CFC-113, held in one bulk tank, and HCFC-22, held in another bulk tank, are pre-blended at the filling plant, then placed in a small (1,000 U.S. Gallon) run tank designed to supply the blend to the pressure-filler (gasser)--allowing the gasser to fill the can in a one-stage operation for CFC-113/HCFC-22 formulas--are any problems caused by distillation in the (variably filled) run tank, with air remaining in the can and contributing to pressure, etc.?

Note: This filling option, which favors the larger filler, conserves CFC-113, compared with using 750-lb drums of that solvent. The bulk price is also more attractive than the drum price, though this is offset to some degree by a larger inventory.

2. Examining the availability of HCFC-142b (CH_3-$CClF_2$) in light of the potential for its greatly increased use. Currently, the sole U.S. manufacturer is Pennwalt Corporation, which uses a complex, somewhat antiquated synthesis that is reflected in the present $2.40/lb price. Allied-Signal Corporation has made this product in the past at their Baton Rouge, LA facility, but their equipment may not still be in place. DuPont feels HCFC-142b is a major solution to the CFC problem, has developed a much-improved manufacturing process, and promises to supply HCFC-142b at a lower cost.

LONGER-RANGE CFC REDUCTIONS

Longer-range CFC reductions will probably be associated with eliminating the use of CFC-113, as in the following:

- Elimination of CFCs (as CFC-113) from the still-permitted "essential-use," non-medical products; and

- Elimination of CFCs from metered-dose inhalant sprays, hair restorers, and contraceptive foam products which are under FDA control and the subject of NDAs.

Subject to laboratory confirmations and field testing, one or two of the "future alternative" propellants, namely HCFC-141b and HCFC-123, might be used to replace CFC-113. Because of the slight flammability of HCFC-141b, it may be necessary to blend it with nonflammable HCFC-123. The proportion of HCFC-123 required to produce nonflammability is not known.

Based on 1988 production figures, and conforming with the suggested rule earlier in this section, a projection of about 7.55 million pounds of CFC-113 a year must be considered. (Note: This is the ozone depletion equivalent of about 40.25 million pounds of 1,1,1-trichloroethane - inhibited, of which about 6,000 million pounds are said to be made annually.)

Replacement formulations have been suggested earlier in Section 4: in particular, Formulas XXV and XXVII for the microcrystalline suspension type metered-dose inhalant drugs. Little or no work toward commercializing these options has yet been done by the pharmaceutical industry--either in the U.S. or worldwide. They require three propellants not now in commercial production, although each is available from industry pilot plant production.

Industry spokesmen suggest they _may_ have formulations available in two to three years, after which they could go to the FDA to begin the 3- to 5-year NDA process. By this time, the propellants will be commercially available. In summary, they are visualizing a commercialization date of 1996-1999.

One of the two "best formula" options, Formula XXVII, requires 4.5% CFC-113 as part of the slurrying agent for nonflammability. This would amount to a use-level of about 0.176 million pounds per year, based on present U.S. production volumes.

Formula XXVII could be a reserve option: if the toxicological results obtained for HCFC-123 preclude this propellant from being used in inhalant sprays, the alternative Formula XXVII would have to be considered. The other option would be to force the pharmaceutical industry to somehow cope with the challenge of slurrying and grinding their drug/excipient items with a liquid, flammable propellant before filling aerosol canisters with the mixture. The industry states that they do not possess this technology, although the possibility of cryogenic slurrying exists.

The other medical products, solution-type inhalants, the virucidal inhalant for bronchopneumonia victims, contraceptive foam, and hair restorer, make up approximately 10-12% of the CFCs used for medical purposes, or some 0.4 million pounds a year. This industry should be amenable to the commercialization date of 1996-1999 mentioned earlier for the microcrystalline suspension inhalant sprays.

SUMMARY

Several possible interim alternative formulations for the seven categories of CFC aerosols still considered "essential" have been discussed in detail. Elements of the proposed CFC reduction plan have also been discussed and compared with the Montreal Protocol for the years 1989-2000. Further recommended studies on aerosol formulations are also discussed.

A longer-term CFC reduction plan would first involve eliminating CFC-113 for all non-medical aerosol products (assuming "future alternatives" are useful) and would effect a reduction of 84.3%--from 25.5 million pounds a year of CFCs used to 4.0 million pounds a year. This could occur by 1993 or 1994.
The final phase would consist of eliminating CFCs from the "essential-use" medical aerosol products by between 1996 and 1999.

References

1.   43 FR 11301: March 17, 1978.

2.   40 CFR Part 82, (30566-30602) August 12, 1988.

3.   United Nations Environment Programme, Montreal Protocol on Substances that Deplete the Ozone Layer, Final Act, 1987.

4.   1,1,1-trichloroethane (methyl chloroform) is also an ozone-depleting substance, although it is not subject to the Montreal Protocol.

5.   Dunn, D. P. "CFC Propellants Today," Part 1, Aerosol Age, July 1988.

6.   The information in the "Notes" to Table 3 was provided by John J. Daly. Jr. of DuPont, February 1989.

7.   Communication with Trevor Lloyd, McLaughlin-Gormley-King Co., February 1989.

8.   Communication with Carl Olson, Technical Products Corp., February 1989.

9.   This was the conclusion reached by the Scientific Working Group at the third session of the ad hoc Working Group of Legal and Technical Experts for the Preparation of a Protocol on CFCs to the Vienna Convention for the Protection of the Ozone Layer, Geneva, Switzerland, April 27, 1987 (UNEP/WG.172/CRP.9).

Appendix A—Additional Information on MDIDs

The addendum in the following pages provides selected information that may be of interest concerning meter-spray CFC-containing ethical drug aerosol products.

ALUPENT METERED DOSE INHALER (ORAL) Boehringer-Ingelheim Pharmaceuticals, Inc.

Adrenergic bronchodilator.

0.225 g Meta-proterenol Sulfate, USP in a 15 mL container

Package provides 300 inhalations of 0.65 mg per dose. Two to three doses = equivalent to one inhalation. Limit: 12 inhalations per day.

Excipient: Sorbitan Trioleate

AZMACORT INHALER (ORAL)   William H. Rorer, Inc.

Anti-inflammatory Corticosteroid--for control of bronchial asthma.

0.060 g Triamcinolone Acetonide in a 20 g canister.

Package provides at least 240 inhalations of about 200 micrograms (mcg) per dose, of which 100 mcg are delivered from the unit--in vitro. (Above 240 inhalations, the amount delivered may be inconsistent.)

Formula: Microcrystalline suspension of drug in 1% ethanol, plus CFC-
           12. (Suggest a slurry of 23.08% drug in ethanol is prepared.)

Toxicology: Teratogenic to rats and rabbits, causing cleft palate
              and/or internal hydrocephaly and/or axial skeletal
              defects at low incidence. Typical findings for glucocor-
              ticoids in animals.

AEROBID INHALER SYSTEM (ORAL)   Forest Pharmaceuticals, Inc.

Anti-inflammatory and anti-allergic corticosteroid--for bronchial asthma.

About 0.027 g Flunisolide in a 7 g canister.

Package provides 100 inhalations of about 250 mcg drug, of which total systemic availability is about 40%--at 2.0 mg per day. The 2.0 g/day level is the chronic adult maximum.

Formula: Microcrystalline suspension of Flunisolide hemihydrate, with sorbitan trioleate (dispersant) and CFC-11, CFC-12, and CFC-114.

Toxicology: Teratogenic to rats and rabbits at 40 mcg/kg/day to 200 mcg/kg/day, as are other corticosteroids. Also feto-toxic.

BECLOVENT ORAL INHALER (ORAL)   Glaxo, Inc. (Research Triangle Park, NC)

A corticosteroid for control of bronchial asthma.

About 0.0084 g of Beclomethasone Dipropionate, USP in a 16.8 g canister.

Package provides at least 200 inhalations. For adults, the maximum daily intake should not exceed 20 inhalations.

Formula: Microcrystalline suspension of beclomethacone dipropionate-trichloromonofluoromethane clathrate in oleic acid and CFC-11 plus CFC-12.

Toxicology: Teratogenic and embryocidal in the mouse and rabbit (but not the rat) when applied at ten times the maximum human dose per kg; e.g., cleft palate and absence of tongue.

BECONASE NASAL INHALER (NASAL)   Glaxo, Inc. (Research Triangle Park, NC)

Identical to Beclovent Oral Inhaler, except for nosepiece.

DECADRON PHOSPHATE RESPIHALER (ORAL)   Merck, Sharp & Dohme Division

Adrenocorticosteroid--for treatment of bronchial asthma.

About 0.107 g of Dexamethasone Phosphate in a package provides about 170 inhalations in a 12.6 g canister.

Formula: Suspension of 0.5 - 4-micron particles of drug in 2% ethanol and fluorocarbons.

Toxicology:

DECADRON PHOSPHATE TURBINAIRE (NASAL)   Merck, Sharp & Dohme Division

Identical to Decadron Phosphate Respihaler except for nosepiece.

MEDIHALER ERGOTAMINE (ORAL)   Riker Laboratories, Inc.

Blood vessel constrictor (cranial)--for treatment of migraines and prodrome.

About 0.225 g of Ergotamine Tartrate in a 2.5 mL (about 3.45 g) vial.

P.36 mg drug per dose; thus, about 25 inhalations per vial.

Formula: Fine particle suspension of ergotamine tartrate, with sorbitan
          trioleate, CFC-11, CFC-12, and CFC-114. Drug is about 0.65%
          w/w.

Toxicology:

MEDIHALER-ISO (ORAL)   Riker Laboratories, Inc.

Adrenergic bronchodilator--for control of bronchial asthma.

About 0.030 g of Isoproterenol Sulfate in a 15 mL (20.4 g) vial.

0.08 mg of drug per dose, and 300 doses per vial.

Formula: A fine suspension of about 0.15% w/w Isoproterenol Sulfate
          powder in sorbitan trioleate, CFC-11, CFC-12, and CFC-114.

Toxicology:

MEDIHALER-EPI (ORAL)   Riker Laboratories, Inc.

Adrenergic bronchodilator--for temporary relief from bronchial asthma.

Epinephrine Bitartrate in a 15 mL (20.4 g) vial.

Each inhalation delivers 0.3 mg of the drug.

Formula: Epinephrine Bitartrate (suspension?) in cetylpyridinium
          chloride, sorbitan trioleate, CFC-11, CFC-12, and CFC-114.

DUO-MEDIHALER (ORAL)   Riker Laboratories, Inc.

Adrenergic bronchodilator--for control of bronchial asthma.

Isoproterenol HCl and Phenylephrine Bitartrate in 15 mL and 22.5 mL
vials.

Each use releases 0.16 mg Isoproterenol HCl and 0.24 mg Phenylephrine
Bitartrate in about 0.05 mL of inerts.

The 15 mL vial provides 300 inhalations.

Formula: Two drugs in micronized particles suspended in cetylpyridinium
          chloride, sorbitan trioleate, CFC-11, CFC-12, and CFC-114.

Toxicology:

OTRIVIN (NASAL)   Geigy Pharmaceuticals Division

Nasal decongestant.

Xylometazoline Hydrochloride, USP.   (Meter-Spray; OTC.)

0.1% drug, in 15 mL aerosol bottle--plastic coated.

Formula:

Toxicology:

PROVENTIL (ORAL)   Schering Corporation

Beta-adrenergic bronchodilator--for reversible obstructive airway
disease, and for prevention of exercise-induced bronchospasm.

Each use discharges 0.090 mg of Albuterol in about 0.126 g of inerts
(c.a. 0.093 mL minimum). The package size is 17.0 g (12.5 mL).

The canister provides at least 200 doses.

Formula: 0.11% Albuterol, Oleic Acid, CFC-11, and CFC-12.

Toxicology: Teratogenic in mice at 14X the maximum human dose.

VANCENASE NASAL INHALER (NASAL)   Schering Corporation

A glucocorticosteroid--relief of symptoms of rhinitis and inflammations.

One inhalation provides 0.042 mg of the drug. Package size is 16.8 g.

8.4 mg of drug is present in the canister, equaling 0.05% (200 doses/can).

Formulation: Microcrystalline suspension of beclomethasone Dipro-
               pionate trichloromonofluoromethane clathrate in oleic
               acid, CFC-11, and CFC-12.

Toxicology: Teratogenic to laboratory animals.

VANCERIL INHALER (ORAL)   Schering Corporation

Same as the above, except for use of oral applicator.

TORNALATE (ORAL)   Winthrop-Breon Laboratories Div.
                                       Sterling Drug, Inc.
                                       Mfr'd by Sterling Pharmaceuticals, Inc.

Beta adrenergic bronchodilator--for bronchial asthma and bronchospasm.

0.8% Bitolterol Mesylate, 38% ethanol, Ascorbic Acid, Saccharin,
Menthol, CFC-12, and CFC-114.

Bottle provides at least 300 doses of about 0.050 mL each, containing
0.37 mg of the drug.

Teratology: Oral doses to rats and rabbits up to 557 times the
               maximum human inhalation dose, and in rats to 284 times
               that dose, produce no teratologic effects. (Some cleft
               palates were obtained through subcutaneous injection.)

VENTOLIN INHALER (ORAL)   Glaxo, Inc. (Research Triangle Park)

Selective Beta$_2$Adrenergic Bronchodilator--relief of bronchospasm.

About 0.118% Albuterol in a 17.0 g canister, providing over 200 inhalations of 90 mcg drug each.

Formulation: Drug (micronized suspension), in oleic acid, CFC-11, and
                CFC-12.

Toxicology: Teratogen, especially via subcutaneous routes.

BRETHAIRE (ORAL)   Geigy Pharmaceuticals Division

Beta-Adrenergic Bronchodilator--relief of bronchospasm.

Each 7.5 mL (10.5 g) canister provides about 300 x 0.25 mg inhalations.

Formulation:

	g/10.5 g Can	Percentage (w/w)
Terbutaline Sulfate	0.075	0.714
Sorbitan Trioleate	0.105	1.000
CFC-11	2.580	24.571
CFC-114	2.580	24.571
CFC-12	5.160	49.144

Toxicology: None.

SUMMARY

BRANDNAME	MARKETER	TYPE ACTION
Alupent Metered Dose Inhaler	Boehringer-Ingelheim	ORAL Bronchodilator
Aerobid Inhaler System	Forest Pharmaceutical	ORAL Corticosteroid
Azmacort Inhaler	William H. Rorer, Inc.	ORAL Corticosteroid
Beclovent Oral Inhaler	Glaxo, Inc.	ORAL Corticosteroid
Beconase Nasal Inhaler	Glaxo, Inc.	NASAL Corticosteroid
Brethaire	Geigy Pharmaceuticals	ORAL Bronchodilator
Decadron Phosphate Respihaler	Merck, Sharp & Dohme	ORAL Corticosteroid
Decadron Phosphate Turninaire	Merck, Sharp & Dohme	NASAL Corticosteroid
Duo-Medihaler	Riker Laboratories, Inc.	ORAL Bronchodilator
Medihaler - Epi	Riker Laboratories, Inc.	ORAL Bronchodilator
Medihaler - Iso	Riker Laboratories, Inc.	ORAL Bronchodilator
Medihaler Ergotamine	Riker Laboratories, Inc.	ORAL CONSTRICTOR
Proventil	Schering Corporation	ORAL Bronchodilator
Tornalate	Winthrop-Breon Labs	ORAL Bronchodilator
Vancenase Nasal Inhaler	Schering Corporation	NASAL Corticosteroid
Vanceril Inhaler	Schering Corporation	ORAL Corticosteroid
Ventolin Inhaler	Glaxo, Inc.	ORAL Bronchodilator

PRODUCT DISTRIBUTIONS

BRANDNAME OF INHALERS	NET WEIGHT (g)	BRONCHODILATOR ORAL	CORTICOSTEROID ORAL	CORTICOSTEROID NASAL	ERGOT. TARTR. ORAL
Alupent	21	XX			
Azmacort	20		XX		
Aerobid	7		XX		
Beclovent	17		XX		
Beconase	17			XX	
Decadron Ph. Resp.	13		XX		
Decadron Ph. Turb.	13			XX	
Brethaire	11	XX			
Medihaler - Epi	20	XX			
Medihaler - Iso	20	XX			
Medihaler - Ergot.	3.5				XX
Duo-Medihaler	20 & 30	XX			
Proventil	17	XX			
Tornalate	17	XX			
Vancenase	17			XX	
Vanceril	17		XX		
Ventolin	17	XX			
	16.0	8	5	3	1

Appendix B—DOT Regulations for Compressed Gases

Department of Transportation
Regulations for Compressed Gases
Title 49 Code of Federal Regulations
Part 173

(d) *Poisonous mixtures.* A mixture containing any poisonous material, Class A, or irritating material in such proportions that the mixture would be classed as poisonous under § 173.326(a) or § 173.381(a) must be shipped in packagings as authorized for these poisonous materials.

[29 FR 18743, Dec. 29, 1964. Redesignated at 32 FR 5606, Apr. 5, 1967, and amended by Amdt. 173-70, 38 FR 5309, Feb. 27, 1973, Amdt. 173-94, 41 FR 16079, Apr. 15, 1976; 45 FR 32697, May 19, 1980]

§ 173.306 Limited quantities of compressed gases.

(a) Limited quantities of compressed gases for which exceptions are permitted as noted by reference to this section in § 172.101 of this subchapter are excepted from labeling (except when offered for transportation by air) and, unless required as a condition of the exception, specification packaging requirements of this subchapter when packed in accordance with the following paragraphs. In addition, shipments are not subject to Subpart F of Part 172 of this subchapter, to Part 174 of this subchapter except § 174.24 and to Part 177 of this subchapter except § 177.817.

(1) When in containers of not more than 4 fluid ounces capacity (7.22 cubic inches or less) except cigarette lighters. Special exceptions for shipment of certain compressed gases in the ORM-D class are provided in Subpart N of this part.

(2) When in metal containers filled with a material that is not classed as a hazardous material to not more than 90 percent of capacity at 70° F. then charged with nonflammable, nonliquefied gas. Each container must be tested to three times the pressure at 70° F. and, when refilled, be retested to three times the pressure of the gas at 70° F. Also, one of the following conditions must be met:

(i) Container is not over 1 quart capacity and charged to not more than 170 psig at 70° F. and must be packed in a strong outside packaging, or

(ii) Container is not over 30 gallons capacity and charged to not more than 75 psig at 70° F.

(3) When in a metal container charged with a solution of materials and compressed gas or gases which is nonpoisonous, provided all of the following conditions are met. Special exceptions for shipment of aerosols in the ORM-D class are provided in Subpart N of this part.

(i) Capacity must not exceed 50 cubic inches (27.7 fluid ounces).

(ii) Pressure in the container must not exceed 180 psig at 130° F. If the pressure exceeds 140 psig at 130° F., but does not exceed 160 psig at 130° F., a specification DOT 2P (§ 178.33 of this subchapter) inside metal container must be used; if the pressure exceeds 160 psig at 130° F., a specification DOT 2Q (§ 178.33a of this subchapter) inside metal container must be used. In any event, the metal container must be capable of withstanding without bursting a pressure of one and one-half times the equilibrium pressure of the content at 130° F.

(iii) Liquid content of the material and gas must not completely fill the container at 130° F.

(iv) The container must be packed in strong outside packagings.

(v) Each completed container filled for shipment must have been heated until the pressure in the container is equivalent to the equilibrium pressure of the content at 130° F. (55° C.) without evidence of leakage, distortion, or other defect.

(vi) Each outside packaging must be marked "INSIDE CONTAINERS COMPLY WITH PRESCRIBED REGULATIONS."

(b) *Exemptions for foodstuffs, soap, biologicals, electronic tubes, and audible fire alarm systems.* Limited quantities of compressed gases, (except poisonous gases as defined by § 173.326) for which exceptions are provided as indicated by reference to this section in § 172.101 of this subchapter, when in accordance with one of the following paragraphs are excepted from labeling (except when offered for transportation by air) and the specification packaging requirements of this subchapter. In addition, shipments are not subject to Subpart F of Part 172 of this subchapter, to Part 174 of this subchapter except § 174.24 and to Part 177 of this subchapter, except § 177.817. Special exceptions for shipment of certain compressed gases in

144   Alternative Formulations and Packaging to Reduce Use of CFCs

§ 173.306

49 CFR Ch. I (10-1-89 Edition)

the ORM-D class are provided in Subpart N of this part.

(1) Foodstuffs or soaps in a nonrefillable metal container not exceeding 50 cubic inches capacity (27.7 fluid ounces), with soluble or emulsified compressed gas, provided the pressure in the container does not exceed 140 p.s.i.g. at 130° F. The metal container must be capable of withstanding without bursting a pressure of one and one-half times the equilibrium pressure of the content at 130° F.

(i) Containers must be packed in strong outside packagings.

(ii) Liquid content of the material and the gas must not completely fill the container at 130° F.

(iii) Each outside packaging must be marked "INSIDE CONTAINERS COMPLY WITH PRESCRIBED REGULATIONS."

(2) Cream in refillable metal receptacles with soluble or emulsified compressed gas. Containers must be of such design that they will hold pressure without permanent deformation up to 375 psig and must be equipped with a device designed so as to release pressure without bursting of the container or dangerous projection of its parts at higher pressures. This exception applies to shipments offered for transportation by refrigerated motor vehicles only.

(3) Nonrefillable metal containers charged with a solution containing biological products or a medical preparation which could be deteriorated by heat, and compressed gas or gases, which is nonpoisonous and nonflammable. The capacity of each container may not exceed 35 cubic inches (19.3 fluid ounces). The pressure in the container may not exceed 140 psig at 130° F., and the liquid content of the product and gas must not completely fill the containers at 130° F. One completed container out of each lot of 500 or less, filled for shipment, must be heated, until the pressure in the container is equivalent to equilibrium pressure of the content at 130° F. There must be no evidence of leakage, distortion, or other defect. Container must be packed in strong outside packagings.

(4) Electronic tubes, each having a volume of not more than 30 cubic inches and charged with gas to a pressure of not more than 35 psig and packed in strong outside packagings.

(5) Audible fire alarm systems powered by a compressed gas contained in an inside metal container when shipped under the following conditions:

(i) Each inside container must have contents which are not flammable, poisonous, or corrosive as defined under this part,

(ii) Each inside container may not have a capacity exceeding 35 cubic inches (19.3 fluid ounces),

(iii) Each inside container may not have a pressure exceeding 70 psig at 70° F. and the liquid portion of the gas may not completely fill the inside container at 130° F., and

(iv) Each nonrefillable inside container must be designed and fabricated with a burst pressure of not less than four times its charged pressure at 130° F. Each refillable inside container must be designed and fabricated with a burst pressure of not less than five times its charged pressure at 130° F.

(c) *Fire extinguishers.* Fire extinguishers charged with limited quantities of a compressed gas to not more than 240 psig at 70° F. are excepted from labeling (except when offered for transportation by air) and the specification packaging requirements of this subchapter when shipped under the following conditions. In addition, shipments are not subject to Subpart F of Part 172 of this subchapter, to Part 174 of this subchapter except § 174.24 and to Part 177 of this subchapter except § 177.817.

(1) Each fire extinguisher must be shipped as an inside packaging;

(2) Each fire extinguisher must have contents which are not flammable, poisonous, or corrosive as defined under this part;

(3) Each fire extinguisher under stored pressure may not have an internal volume exceeding 1,100 cubic inches. For fire extinguishers not exceeding 35 cubic inches capacity, the liquid portion of the gas plus any additional liquid or solid must not completely fill the container at 130° F. Fire extinguishers exceeding 35 cubic inches capacity may not contain any liquefied compressed gas;

(4) Each fire extinguisher manufactured on and after January 1, 1976, must be designed and fabricated with a burst pressure of not less than six times its charged pressure at 70° F. when shipped.

(5) Each fire extinguisher must be tested, without evidence of failure or damage, to at least three times its charged pressure at 70° F. but not less than 120 psig before initial shipment. For any subsequent shipment, each fire extinguisher must be in compliance with the retest requirements of the Occupational Safety and Health Administration Regulations of the Department of Labor, 29 CFR 1910.157(e), and;

(6) Each fire extinguisher must be marked to indicate the year of the test (within 90 days of the actual date of the original test) and "MEETS DOT REQUIREMENTS." This marking will be considered a certification that the fire extinguisher was manufactured in accordance with the requirements of this section.

Note: The words "This extinguisher meets all requirements of 49 CFR 173.306" may be displayed in place of "MEETS DOT REQUIREMENTS" on extinguishers manufactured prior to January 1, 1976.

(7) When Specification 2P or 2Q (§§ 178.33, 178.33a of this subchapter) packagings are used, paragraphs (c)(4)–(6) of this section are not applicable provided each packaging meets the requirements of paragraph (a) of this section.

(d) *Truck bodies or trailers on flat cars; automobiles, motorcycles, tractors, or other self-propelled vehicles.* (1) Except as specified in § 173.21, truck bodies or trailers with automatic heating or refrigerating equipment of the gas burning type may be shipped with tanks containing fuel and equipment operating or not operating, when used for the transportation of other freight and loaded on flat cars as part of a joint rail-highway movement. The heating or refrigerating equipment is considered to be a part of the truck body or trailer and is not subject to any other requirements of this subchapter.

(2) Automobiles, motorcycles, tractors, or other self-propelled vehicles equipped with liquefied petroleum gas or other compressed gas fuel tanks, provided such tanks are securely closed, are not subject to any other requirements for transportation by rail or highway. For transportation by water, see §§ 176.905 and 176.78(k) of this subchapter. For transportation by air, the fuel tank must be removed or emptied and securely closed.

(3) A cylinder which is a component part of a passenger restraint system and is installed in a motor vehicle, charged with nonliquefied, nonflammable compressed gas and having no more than two actuating cartridges per valve, is excepted from the requirements of Parts 170-189 of this subchapter except:

(i) Unless otherwise authorized by the Department, each cylinder must be in compliance with one of the cylinder specifications in Part 178 of this subchapter and authorized for use in § 173.302 for the gas it contains;

(ii) Each cylinder must be in compliance with the filling requirements of § 173.301; and

(iii) Each actuating cartridge must be approved in accordance with § 173.86 and meet the definition set forth in § 173.100(w).

(4) A cylinder which is part of a tire inflator system in a motor vehicle, charged with a nonliquefied, nonflammable compressed gas is excepted from the requirements of Parts 170-189 of this subchapter except:

(i) Unless otherwise authorized by the Department, each cylinder must be in compliance with one of the cylinder specifications in Part 178 and authorized for use in § 173.302 for the gas it contains;

(ii) Each cylinder must be in compliance with the filling requirements of § 173.301.

(iii) Each cylinder must be securely installed in the trunk of the motor vehicle and the valve must be protected against accidental discharge.

Note: A cylinder containing a gas generator may be included within the provisions of this exception if the requirements of § 173.34(d) are satisfied.

(e) *Refrigerating machines.* (1) New (unused) refrigerating machines or components thereof are excepted from the specification packaging require-

ORM-D material (see § 173.500) provided that an ORM-D exception is authorized in specific sections applicable to the material, and that it is prepared in accordance with the following paragraphs. (The gross weight of each package must not exceed 65 pounds and each package offered for transportation aboard aircraft must meet the requirements of § 173.6.)

(1) *Flammable Liquids must be:* (i) In inside metal containers, each having a rated capacity of 1 quart or less, packed in strong outside packagings.

(ii) In inside containers, each having a rated capacity of 1 pint or less, packed in strong outside packagings.

(iii) In inside containers, each having a rated capacity of one gallon or less, packed in strong outside packagings. The provisions of this exception apply only if the flash point of the material is 73° F. or higher.

(2) *Corrosive liquids must be:* (i) In bottles, each having a rated capacity of 1 pint or less, each enclosed in a metal can, packed in strong outside packagings.

(ii) In metal or plastic containers, each having a rated capacity of 1 pint or less, packed in strong outside packagings.

(iii) In metal or plastic inside containers, each having a rated capacity of not over 1 quart, packed in strong outside packaging provided the liquid mixture contains 15 percent or less corrosive material and the remainder of the mixture does not meet the definition of a hazardous material as defined in this subchapter. Not authorized for transportation by air.

(3) *Corrosive solids must be:* (i) In earthenware, glass, plastic or paper containers each having a net weight of 5 pounds or less, packed in strong metal, wooden, or fiberboard outside packagings, each having a net weight of 25 pounds or less.

(ii) In metal, rigid fiber, or composition cans or cartons or rigid plastic containers each having a net weight of 10 pounds or less, packed in strong outside packagings each having a net weight of 25 pounds or less.

(iii) In metal, rigid fiber, or composition cans or cartons or rigid plastic containers, each having a rated capac-

ity of not over 20 pounds, overpacked in metal, wooden or fiberboard outside containers not exceeding 50 pounds net weight provided the solid mixture contains 10 percent or less corrosive material and the remainder of the mixture does not meet the definition of a hazardous material as defined in this subchapter.

(4) *Flammable solids* except for charcoal briquettes must be in inside containers each having a net weight of 1 pound or less, packed in strong outside packagings each having a net weight of 25 pounds or less. Charcoal briquettes may be shipped in packagings having a net weight of 65 pounds or less.

(5) *Oxidizers* must be in inside containers each having a rated capacity of 1 pint or less for liquids or a net weight of 1 pound or less for solids, packed in strong outside packaging each having a net weight of 25 pounds or less.

(6) *Organic peroxides* must be: (i) In inside containers which must be securely packed and cushioned with noncombustible cushioning material in strong outside packagings containing not over 1 pint or 1 pound net quantity of the materials. Cushioning is not required when the liquid is contained in strong, securely closed, plastic packagings, not over 1 ounce capacity each, properly packed to prevent leakage or breakage.

(ii) In strong outside packagings of 24 or less inside fiberboard containers, each having 70 or less securely closed tubes having a maximum fluid capacity of ¼-ounce each and securely packed in noncombustible cushioning material. Each fiberboard container may not contain more than 1 pint of liquid.

(7) *Poison B liquids or solids* must be in inside containers, each having a rated capacity of 8 ounces or less by volume for liquids or of 8-ounces or less net weight for solids packed in strong outside packagings.

(8) *Compressed gases* must be: (i) In inside containers, each having a water capacity of 4-fluid ounces or less (7.22 cubic inches or less), packed in strong outside packagings.

(ii) In inside metal container charged with a solution of materials and compressed gas or gases which is nonpoisonous, meeting all of the following:

(A) Capacity may not exceed 50 cubic inches (27.7 fluid ounces);

(B) Pressure in the container may not exceed 180 p.s.i.g. at 130° F. (55° C.). If the pressure exceeds 140 p.s.i.g. at 130° F., (55° C.) but does not exceed 160 p.s.i.g. at 130° F., (55° C.) a specification DOT 2P (§ 178.33 of this subchapter) inside metal container must be used; if the pressure exceeds 160 p.s.i.g. at 130° F., (55° C.), a specification DOT 2Q (§ 178.33a of this subchapter) inside metal container must be used. In any event the metal container must be capable of withstanding, without bursting, a pressure of one and one-half times the equilibrium pressure of the contents at 130° F. (55° C.);

(C) Liquid content of the material and gas not completely fill the container at 130° F. (55° C.);

(D) The containers must be packed in strong outside packagings; and

(E) Each completed container filled for shipment must have been heated until the pressure in the container is equivalent to the equilibrium pressure of the content at 130° F. (55° C.) without evidence of leakage, distortion, or other defect.

(iii) In a non-refillable inside metal container of 50 cubic-inch capacity or less (27.7 fluid ounces), with foodstuffs or soaps and with soluble or emulsified compressed gas, provided the pressure in the container does not exceed 140 p.s.i.g. at 130° F. (55° C.). The metal container must be capable of withstanding, without bursting, a pressure of one and one-half times the equilibrium pressure of the contents at 130° F. (55° C.) and must comply with the following provisions:

(A) Containers must be packed in strong outside packagings, and

(B) Liquid content of the material and gas may not completely fill the container at 130° F. (55° C.).

(iv) In refillable inside metal containers with cream and soluble or emulsified compressed gas packed in strong outside packagings. Containers must be of such design that they will hold pressure without permanent deformation up to 375 p.s.i.g. and must

be equipped with a device designed so as to release pressure without bursting of the container or dangerous projection of its parts at higher pressures.

(v) In non-refillable inside metal containers charged with a solution, containing biological products or a medical preparation which could be deteriorated by heat, and compressed gas or gases which is nonpoisonous and nonflammable. The capacity of each container may not exceed 35 cubic inches (19.3 fluid ounces). The pressure in the container may not exceed 140 p.s.i.g. at 130° F. (55° C.), and the liquid content of the product and gas may not completely fill the container at 130° F. (55° C.). One completed container out of each lot of 500 or less, filled for shipment, must be heated, until the pressure in the container is equivalent to the equilibrium pressure of the content at 130° F. (55° C.). There may be no evidence of leakage, distortion, or other defect. Container must be packed in strong outside packagings.

(vi) In electronic tubes, each having a volume of not more than 30 cubic inches and charged withb 'as to a pressure of not more than 35 p.s.i.g. and packed in strong outside packagings.

(vii) In an inside metal container as a component of an audible fire alarm system powered by a compressed gas meeting the following provisions:

(A) Each inside container must have contents which are not flammable, poisonous, or corrosive as defined under this part;

(B) Each inside container may not have a capacity exceeding 35 cubic inches (19.3 fluid ounces);

(C) Each inside container may not have a pressure exceeding 70 p.s.i.g. at 70° F. (21° C.) and the liquid portion of the gas may not completely fill the inside container at 130° F. (55° C.);

(D) Each inside container must be designed and fabricated with a burst pressure of not less than five times its charged pressure at 130° F. (55° C.); and

(E) Each fire alarm system must be packed in a strong outside packaging.

[Amdt. 173-94, 41 FR 16091, Apr. 15, 1976, as amended by Amdt. 173-94A, 41 FR 40684,

Appendix C—Metric (SI) Conversion Factors

Quantity	To Convert Form	To	Multiply By
Length:	in	cm	2.54
	ft	m	0.3048
Area:	in²	cm²	6.4516
	ft²	m²	0.0929
Volume:	in³	cm³	16.39
	ft³	m³	0.0283
	gal	m³	0.0038
Mass (weight):	lb	kg	0.4536
	oz	kg	0.0283
	short ton (ton)	Mg	0.9072
	short ton (ton)	metric ton (t)	0.9072
Pressure:	atm	kPa	101.3
	mm Hg	kPa	0.133
	psig	kPa	6.895
	psig	kPa*	$((psig)+14.696) \times (6.895)$
Temperature:	°F	°C*	$(5/9) \times (°F-32)$
	°C	K*	$°C+273.15$
Caloric Value:	Btu/lb	kJ/kg	2.326
Enthalpy:	Btu/lbmol	kJ/kgmol	2.326
	kcal/gmol	kJ/kgmol	4.184
Specific-Heat Capacity:	Btu/lb-°F	kJ/kg-°C	4.1868
Density:	lb/ft³	kg/m³	16.02
	lb/gal	kg/m³	119.8
Concentration:	oz/gal	kg/m³	
	quarts/gal	cm³/m³	25.000
Flowrate:	gal/min	m³/min	0.0038
	gal/day	m³/day	0.0038
	ft³/min	m³/min	0.0283
Velocity:	ft/min	m/min	0.3048
Viscosity:	centipoise (CP)	Pa-s (kg/m-s)	0.001

*Calculate as indicated

Part II

Alternative Formulations and Aerosol Dispensing Systems

The information in Part II is from *Aerosol Industry Success in Reducing CFC Propellant Usage,* prepared by Thomas P. Nelson and Sharon L. Wevill of Radian Corporation for the U.S. Environmental Protection Agency, November 1989.

Section I
Alternative Aerosol Formulations

1. Introduction

There is an urgent need to reformulate aerosol products into compositions that no longer contain chlorofluorocarbons ($C_xCl_yF_z$). As early as 1973, scientists recognized that these compounds had very long atmospheric lives and could ultimately penetrate the stratospheric ozone layer at altitudes of between about 14 to 27 km. Once in the stratosphere, CFCs are bombarded with high-energy radiation from the sun, splitting off a chlorine atom that reacts with thousands of ozone molecules and reduces them to ordinary oxygen. Although the ozone is reformed by natural processes over time, the overall effect is of ozone depletion.

During September 1987, a meeting held in Montreal, Canada was attended by representatives of many nations. A treaty known as the Montreal Protocol was developed calling for the orderly reduction of chlorofluorocarbon (CFC) production, roughly according to the following schedule:

By July 1, 1989   Reduction to the 1986 average production level [15-25% actual reduction in the U.S. because of the growth in CFC use since 1986; Ozone Depletion Potential (ODP) basis.]

By July 1, 1993   Reduction to 80% of the 1986 average level, ODP basis.

By July 1, 1998   Reduction to 50% of the 1986 average level, ODP Basis.

151

As of October 1989, the treaty had been ratified by 43 nations plus the European Community (EC) as a bloc, which together produce approximately 90% of the world tonnage of CFCs.

The results of stratospheric studies made after the Montreal Protocol now strongly suggest that the reduction plan is insufficient to prevent a further depletion of ozone.

Another problem has surfaced, however. As CFCs are phased out, they will be replaced by such chemicals as HCFC-22, 1,1,1-trichloroethane (methyl chloroform) and similar substances, many of which can also deplete stratospheric ozone. Table 1 provides comparative figures.

In 1985, HCFC-22 was responsible for only 0.4% of ozone removal, while 1,1,1-trichloroethane caused about 5.1% ozone removal and CFC-12 was responsible for about 40.1% of the total ozone removal caused by the compounds listed in Table 1. Except for the hydrocarbons and nitrogen, all the compounds in Table 1 are anthropogenically produced.

Such compounds as HCFC-123, HCFC-124, HFC-134a, and HCFC-141b are currently undergoing extensive toxicological testing that is expected to continue until about 1992. HCFC-123 currently has an Acceptable Exposure Limit (AEL), or TLV, of 100 ppm, but this may be changed to somewhere in the 50 to 100 ppm range as further results are developed. Similarly, HCFC-141b may get an AEL of 100 to 300 ppm. Results of the Ames Salmonella Test for HCFC-22, HCFC-141b, and HCFC-142b show positive mutagenic results for all the compounds, but extensive animal testing has clouded the meaning of the Ames results.

TABLE 1.   EMISSIONS AND OZONE DEPLETION POTENTIALS OF AEROSOL
PROPELLANTS AND RELATED COMPOUNDS

Compound	Structure	1985 Emissions (k tons/year)	Ozone Depletion Potential (ODP) (CFC-11 = 1)[a]
CFC-11	CCl_3F	281	1.00
CFC-12	CCl_2F_2	307	1.0
CFC-113	$CCl_2F-CClF_2$	138	0.8
CFC-114	$CClF_2-CClF_2$	(low)	0.8
CFC-115	$CClF_2-CF_3$	(very low)	0.4 (0.15)[b]
HCFC-22	$CHClF_2$	72	0.05
HCFC-123	$CHCl_2-CF_3$	0	0.02
HCFC-132b	$CH_2Cl-CClF_2$	0	0.05
HCFC-124	$CHClF-CF_3$	0	0.02
HFC-134a	CH_2F-CF_3	0	0
HFC-125	CHF_2-CF_3	0	0
HCFC-141b	CH_3-CCl_2F	0	0.10
HCFC-142b	CH_3-CClF_2	(low)	0.06
HFC-152a	CH_3-CHF_2	0	0
Halon 1211	$CBrClF_2$	3	2.7
Halon 1301	$CBrF_3$	3	10.0
Halon 2402	$CBrF_2-CBrF_2$	(very low)	5.6
Carbon Tetrachloride	CCl_4	66	1.2
1,1,1-Trichloroethane	CH_3-CCl_3	474	0.10 (0.15)[b]
Hydrocarbons	C_3H_8, etc.	(very large)	0
CO_2, N_2O & N_2	CO_2, N_2O, N_2	(very large)	0[c]
Dimethyl ether	CH_3-O-CH_3	42	0

[a]UNEP Data of 18-OCT-1988.
[b]Isaksen, et al (1988).
[c]N_2O can destroy stratospheric ozone but its ODP is undefined.

Many of the future alternative compounds are nonflammable, while others are flammable. HCFC-123 is nonflammable, but a mixture of this gas and 8.8% isobutane is marginally flammable. HCFC-141b has a flammable range of 6.4 to 15.1%, while HCFC-142b's flammable range is 6.7 to 14.9%. HFC-134a, which is being groomed as a replacement for most uses of CFC-12, is nonflammable. HCFC-22 is the only nonflammable (1), commercially available CFC alternative that the industry will have until about 1993 or 1994, when some or all of the second generation CFC alternatives should come onto the market. It is only marginally nonflammable; the addition of 6% isobutane, or 8.6% ethanol to HCFC-22 will produce mixtures of borderline flammability.

The worldwide aerosol business is highly diversified. In 1989, the U.S. will produce about 3 billion units (95% non-CFC aerosols), or 35% of the world total of about 8.6 billion units. Western Europe will produce about 39%, Japan 5%, Brazil 2%, and Mexico 0.5%. Per capita usage is 11 units per person in the U.S.: the typical home contains 46 aerosol products, averaging 206 g per unit. Since the purchase of aerosols is often discretionary (they are not generally considered to be utility products) the per capita usage in different countries is a reflection of both availability and of the relative standard of living. The more hours a person must work to purchase an aerosol, the fewer will be purchased.

Apart from the usual competitive pressures, there is a strong motivation to reduce the costs of aerosol products in order to increase sales. In the U.S., hydrocarbon propellants cost less than 20% of the rapidly escalating costs of CFCs. They are therefore the propellants of choice unless special properties are required, such as better solvent action or reduced flammability. Approximately 81% of U.S. aerosols are pressurized with propane, n-butane, isobutane, or their blends. Another 7% use carbon dioxide, and the remaining 12% use nitrous oxide, CFCs, dimethyl ether, nitrogen, HFC-152a and HCFCs, in approximately that order. The few CFC aerosols remaining after the general ban on these products was imposed during 1978 are those permitted by exclusion, exemption, or those that are not regulated.

Hydrocarbon propellants are already in wide use throughout the world.
Examples are as follows: United Kingdom, a market share of 30%; West Germany,
80%; Brazil, 88%; Mexico, 92%; and Canada, 78%. The next preferred CFC
alternative is dimethyl ether (DME, or dimethyl oxide). DME alternatives are
about 10% more costly than the hydrocarbon alternatives in Western Europe,
100% more expensive in the U.S., and even more costly, or unavailable, in
other parts of the world. The major producers are Western Europe, with a
capacity of 60,000 tons, Japan, the U.S., Canada, and Australia. Dimethyl
ether is flammable. It is also a very strong solvent, sometimes causing
gasket failures in equipment, aerosol corrosion, valve seal leakage, and
excessive swelling of some elastomers. It is highly water soluble, and can be
used as a way of incorporating water into solution in selected aerosol
products, such as hair sprays and personal deodorants. Table 2 compares the
physical properties of the non-CFC aerosol propellants.

Although carbon dioxide, nitrous oxide, and nitrogen are widely available
throughout the world, they have either been ignored or little used as aerosol
propellants. These gases are inexpensive, but special equipment is often
required to add them to aerosol containers. The simplest of these is the
gasser-shaker, of either in-line or rotary construction, which is shaken at a
preset frequency and amplitude for a fixed period of time. It is connected
through the valve to a supply of gas regulated to a pressure of approximately
142 to 178 psig (10.0 to 12.5 bars). Valve designs are available that will
facilitate gas flow into the can, even with the button attached. Since the
quantity of gas added will be in the range of 3 to 28 g, depending on can size
and content, the weight increase of the dispenser is used as a basis for
machine adjustments. Table 3 shows the potential uses of these propellants
for several representative products.

HCFC-22 is widely used throughout much of the world as a specialty
refrigerant and freezant. Despite its nonflammability (1) and relatively low
price (five times more costly than hydrocarbons, in the U.S.), it is not much
used. It is limited by its high pressure, which makes it necessary to use 40%

TABLE 2.   PHYSICAL PROPERTIES OF NON-CFC AEROSOL PROPELLANTS

Product	Formula	Boiling Point (°C)	Vapor Pressure (bar) 21°C	55°C	Density at 21° (g/mL)	Flammable Range v.%
nbutane	$n.C_4H_{10}$	-2	1.20	4.79	0.580	1.8 - 8.6
isobutane	$i.C_4H_{10}$	-11	2.17	7.02	0.559	1.8 - 8.5
Propane	C_3H_8	-42	7.60	18.17	0.503	2.2 - 9.5
Dimethyl Ether	$(CH_3)_2O$	-25	4.43	12.40	0.661	3.3 - 18.0
HCFC-22	$CHClF_2$	-41	8.52	20.92	1.208	0
HCFC-142b	CH_3-CClF_2	-10	2.04	6.87	1.123	6.7 - 14.9
HFC-152a	CH_3-CHF_2	-25	4.42	12.36	0.911	3.9 - 16.9
Carbon Dioxide	CO_2	-78	58.45	N/A	0.721	0
Nitrous Oxide	N_2O	-88	52.47	N/A	0.718	0
Nitrogen	N_2	-155	N/A	N/A	N/A	0

FUTURE PROPELLANTS

Product	Formula	Boiling Point (°C)	Vapor Pressure (bar) 21°C	55°C	Density at 21° (g/mL)	Flammable Range v.%
HCFC-123	$CHCl_2-CF_3$	28	-0.2	1.7	1.470	0
HCFC-124	$CHClF-CF_3$	-11	3.22	8.8	1.368	0
HFC-125	CHF_2-CF_3	-95	N/A	N/A	N/A	0
HFC-134a	CH_2F-CF_3	-32	5.47	14.3	1.203	0
HCFC-141b	CH_3-CCl_2F	33	-0.3	1.2	1.231	6.4 - 15.1

N/A = Non Applicable, above Critical Temperature.

TABLE 3.   PRODUCT APPLICATIONS OF CARBON DIOXIDE, NITROUS OXIDE, AND NITROGEN

__Carbon Dioxide__

Hydroalcoholic disinfectant/deodorant sprays.
Bug killers:
      Ant and roach killers
      Wasp and hornet killers
Lubricants.
Anti-statics, soil repellants, and wrinkle removers for textiles.

__Nitrous Oxide__

Whipped creams.
Heavy-texture specialty foams.
Windshield and car lock de-icer sprays.
Furniture polish.

__Nitrogen__

Sterile saline solutions for rinsing contact lenses.
Long-range, stream-type wasp and hornet killers.
Injector-type engine cleaners.
Over-pressurant for selected meter-sprayed vitamins and drugs.

or less in formulas and to include suppressive solvents or other propellants to keep the aerosol pressure from being excessive. An interesting blend of HCFC-22/HCFC-142b (40:60) is nonflammable and has a pressure of 63 psig at 70°F (4.43 bar at 21°C). It has been commercialized for perfumes and colognes. HCFC-22 is a good solvent. At less than 28% propellant, its ethanol solutions are lower in pressure than those of CFC-12 and ethanol.

HCFC-142b is used in a few applications in the U.S. and is presently unavailable elsewhere. It is now made by only one supplier, although a second supply source is being developed. As the methyl homolog of HCFC-22, it has many properties in common with the parent compound, except the high pressure. It is more than 12 times as costly as hydrocarbon propellants in the U.S., which has restricted its aerosol applications.

HFC-152a is close to an ideal propellant, except that it is flammable. It is less flammable than hydrocarbon gases, however, and it has typically been used with 70% A-46 (20 mol % propane and 80 mol % isobutane) to produce a propellant for shave creams, depilatories, and mousse products whose foam surface will not momentarily flash if a lighted match is touched to it. The composition is as follows:

60.9% Isobutane
9.1% Propane
30.0% HFC-152a

Since the pressure of the aerosol is about 154 psig at 130°F (11.0 bar at 55°C), according to the partial pressure of remaining air, an extra-strength can is needed.

HFC-152a is noted for its exceptionally low odor and good solvency. It is used to make less flammable colognes and perfumes, especially for those essential oils that might eventually precipitate high-molecular weight resins, fonds, or substantives in the usual ethanol/hydrocarbon (or pure hydrocarbon) systems. Finally, it can be used with many surfactant systems, to partly destabilize aerosol foams, permitting them to be more readily rubbed out on surfaces and not resist liquefaction. A typical product that uses this property is baby oil mousse, which contains 20 to 30% mineral oil.

In the U.S., since HFC-152a is approximately eight times the cost of hydrocarbon propellants, the amounts used in formulas are generally in the 2 to 10% range. It is available in the U.S. and Western Europe, and suppliers claim that distribution systems will be set up to greatly increase world access to this propellant and to HCFC-142b.

The future "CFC alternative" propellants identified in Table 2 are presently undergoing acute, sub-chronic, and chronic (lifetime) toxicological testing. To date, the results have shown some variation in relative toxicity, but indications are that all five compounds will probably be approved for commercial use. The official toxicological reports will be issued in 1992 and 1993, but plans are now in motion to build production facilities well before that time.

In the U.S., DuPont has announced that an existing commercial plant is being converted to produce HCFC-141b and HCFC-142b in 1989. A new plant has been approved to produce large quantities of HFC-134a by 1990. Large quantities of HCFC-123 are already available as a co-product from an existing DuPont facility. And during 1988 DuPont was issued a U.S. Patent on new technology aimed at coproducing HCFC-123 and HCFC-124 in a single process. No schedules for HCFC-124 production have been published.

Other CFC suppliers in the U.S., Western Europe, Japan, and other parts of the world are also studying their options for phasing out CFCs and commercializing various alternatives. The major alternative will probably be HFC-134a, since it will be used to replace CFC-12 in refrigeration, freezant, and air conditioning systems.

An accelerated CFC phase-down program, which goes beyond the Montreal Protocol and is now supported by numerous countries, is based on rapid commercialization and application of the HCFC and HFC alternatives. The science centers around minimizing further increases in the chlorine content of the stratospheric ozone layer.

Table 4 lists the aerosol products currently exempted or excluded from
the general regulatory bans in the U.S. on CFCs for aerosol uses. They serve
life-saving or other medical purposes, or are considered "essential for other
reasons."

A few of these products have been discontinued, such as the drain openers
and small-size tobacco barn sprays. The largest users of CFCs are the mold
release agents, lubricants, and meter-spray inhalant drug products, except for
CFC-12 and CFC-114 small refrigerant recharge units, which many people do not
consider to be true aerosol products.

When considering propellants or propellant/solvent combinations that may
be used for reformulating CFC aerosols, a large number of attributes must be
evaluated. Flammability, toxicology, solvency, cost, availability, solvate
formation, solvolytic stability, dispersancy, pressure, and compatibility are
some of the more essential characteristics. In the late 1980s, a growing
intolerance developed towards propellants and other chemicals that have even
slight effects on the stratospheric and tropospheric ecosystems, that have
greater perceived toxicity than alternatives, or do not degrade in landfills.

TABLE 4.   EXEMPTED, EXCLUDED, OR NONREGULATED CFC AEROSOL PRODUCTS (U.S.)

Mold release agents -- for molds making rubber and plastic items
Lubricants for use on electric or electronic equipment
Lubricants for rotary pill and tablet making presses
Solvent dusters, flushers, degreasers and coatings for electric or electronic
  equipment
Meter-spray inhalant drugs:
    a.   Adrenergic bronchodilators
    b.   Cortico steroids
    c.   Vaso-constrictors - ergotamine tartrate type
Contraceptive vaginal foams - for human use
Mercaptan (as ethyl thiol) mine warning devices
Intruder audio-alarm system canisters - for house and car uses
Flying insect sprays:
    a.   For commercial food-handling areas
    b.   For commercial (international) aircraft - cabin sprays
    c.   For tobacco barns
    d.   For military uses
Military aircraft operational and maintenance uses
Diamond grit abrasive uses
For uses relating to national military preparedness
CFC-115 as a puffing (foaming) agent in certain food aerosols
Automobile tire inflators
Polyurethane foam aerosols
Chewing gum removers
Drain openers
Medical chillers - for localized operations
Medical solvents - as a spray bandage remover
Dusters for non electric or electronic uses - for phonograph records and
      computer tapes
Cleaners for microscope slides and related objects
Foam, whip, or mousse products in general
Small refill units for refrigeration or air-conditioning systems
All other 100% CFC product applications

2. Formulation Guidelines

Dispersancy, one major attribute of aerosol propellants, is the efficiency with which a propellant can produce a fine spray or an acceptable foam. This is illustrated in Table 5.

The dispersancy of blends can be readily calculated. For example, Propellant A-46 (20 mol% propane and 80 mol% isobutane) has a dispersancy of [549 X .2 + 415 X .8] = 442 mL/g at 21.1°C.

A shave cream or mousse, made using either 8% CFC-12/114, 4% A-46, or 2% nitrous oxide will all show the same properties of foam density and overrun. (However, the nitrous oxide formula will have a very high pressure, which can be expected to decrease significantly with use.)

In the years before the CFC aerosol ban of 1978 in the U.S., hair sprays were commonly formulated with 45% CFC-12/11 (55:45), or 40% Propellant A (10% Isobutane, 45% CFC-12, and 45% CFC-11). They are now formulated with 20 to 26% isobutane, sometimes with a small amount or propane added. These examples show the importance of dispersive effect to propellant volume.

The dispersive effect is not linear but is modified by vapor-pressure, solubility factors, and even by the pressure itself. It normally can be used as a general guideline to determine equivalencies when changing from one propellant choice to another.

162

TABLE 5.   DISPERSANCY CHARACTERISTICS OF VARIOUS PROPELLANTS
(In order of Vapor Volume in mL/g)

Propellant	Vapor Volume (mL/g 21.1°C)	Vapor Volume (mL/mL 21.1°C)
Nitrogen	862	N/A
Carbon Dioxide	549	N/A
Nitrous Oxide	549	N/A
Propane	549	280
Dimethyl Ether	523	345
isobutane	415	234
nbutane	415	239
HFC-152a	365	333
HCFC-22	279	337
CFC-115	256	(not available)
HCFC-142b	240	269
HFC-134a	236	283
HCFC-141b	206	253
CFC-12	200	265
CFC-125	198	227
CFC-11	176	261
HCFC-124	176	242
HCFC-123	158	232
CFC-114	141	207
FC-C318	119	179

Note: These propellants boil at <21.2°C (Range: 23° to 33°C.)
N/A — Not Applicable

The aerosol formulator will also have to determine such things as company policy, availability of equipment, and the safety features of the workplace.

Nonflammable propellants (apart from CFCs) consist of nitrogen, nitrous oxide, carbon dioxide, HCFC-22, and a few blends of other propellants with HCFC-22. Future nonflammable propellants will consist of HCFC-123, HCFC-124, HCFC-125, and HFC-134a. Of these, HFC-134a may become available most quickly worldwide. The cost of HCFC and HFC propellants is expected to be about twenty times that of purified hydrocarbons by 1993 or 1994; this may limit their application to relatively specialized products, for example, to perfume meter-sprays in container sizes of 50 mL or less.

When flammable propellants are considered to be within the scope of company operations, the most reasonable choices are isobutane and propane. In some parts of the world the "natural blend" must be used. A typical natural blend will consist of 60% nbutane, 20% isobutane, and 20% propane. It is a broad distillation cut from the gas wells after de-ethanization and partial de-propanization. In some areas, the hydrocarbon may contain large amounts of other impurities. Some gas wells in Canada were found to contain over 50% hydrogen sulfide and alkyl mercaptans (thiols), causing their closure. Wells in Trinidad typically contain 12% unsaturates, such as propylene and isobutylene, making them marginally useful for aerosol applications. Propane/butanes from gas wells in Brazil contain 2.5 to 5.5% unsaturates.

Any contract filler or self-filler contemplating a change from CFC to hydrocarbon propellants should thoroughly investigate such things as availability, purity, fire and building codes or regulations, the cost of conversion, such as the construction of an outside gas house, safety equipment, and electrical revisions. The product development and quality control laboratories should be equipped with explosion-proof hoods, ventilation, and other safety equipment.

When available, dimethyl ether offers a relatively inexpensive alternative to the hydrocarbon propellants. It does not have the potential problem of odor. It is less flammable (on an absolute, LEL, or other scale) but it is also a very strong solvent.

The flammable HCFCs and HFCs are final options, but because of their relatively high cost they may have a minor effect on the worldwide aerosol industry.

Concentrates

Most concentrates are available in the form of suggested formulations by ingredient suppliers. They may be made especially for aerosol uses, or they may be adaptable to aerosol applications. Some, like most paint products, have to be drastically altered before they will work for aerosols.

A large collection of supplier samples and literature is a requisite of any formulating laboratory. The literature should cover properties, uses, compounding techniques, toxicological data and suggested prototype or starting formulations. (Sometimes these formulas have somewhat more of the supplier's product than is really needed.)

After a concentrate has been tentatively developed, there remains the process of adding the correct type and amount of propellant, and using an aerosol valve that will develop the desired spray pattern or foam puff. One of the most important characteristics that the formulator looks for is particle size distribution, which can be of paramount importance. If the droplet size is too coarse, it can be decreased by one of the following techniques:

- Increase the percentage of propellant;

- Increase propellant pressure and/or dispersancy;

- Use a vapor-tap valve or a larger vapor-tap orifice;

- Use a mechanical break-up button;

- Add a low-boiling (volatile, easy breakup) solvent; and

- Reduce the quantity of polymers, thickeners, resins, adhesives, and water.

Approximately 40-50% of the world's 8 billion aerosol products use vapor-tap valves. Such valves have an orifice extending through the side or bottom wall of the valve body and into the head space area. When the orifice of a vapor-tap valve is enlarged to decrease particle size, a price is paid. The negative effects are listed below:

- A broader particle size distribution will generally result.

- A gradual coarsening of the spray may occur during use.

- The internal pressure will decrease, as air and the more volatile propellant ingredients preferentially escape through the vapor-tap orifice.

- The delivery rate will always be lower than without a vapor-tap, and will decrease during use, because of pressure reduction.

The potential problems with vapor-tap valves can be minimized by the following techniques:

- Use the smallest vapor-tap hole that will suffice (a 0.25 mm size may be a good starting point).

- Use a fairly large to large amount of propellant that disperses well (reservoir effect).

- Use a pure propellant; otherwise, the more volatile ingredient will be preferentially discharged, causing a pressure drop.

- Use reasonably large liquid orifices.

- Emphasize any or all of the above in taller cans, since (near emptiness) a liquid column of 150 - 250 mm will have to be maintained in the dip tube just to bring the product into the valve chamber. A greater dynamic pressure potential is needed, compared with shorter can sizes.

As a rule, thin or driving sprays, or sprays with high delivery rates, will be perceived by consumers as "wet" or "cold," although they may be anhydrous. Wet sprays are usually disliked, except for the coating of inanimate surfaces (such as a paint spray or bug killer); they are most disliked for cosmetic items designed to be sprayed on the skin, such as underarm antiperspirants or deodorants. The aerosol antiperspirant provides an interesting challenge because large valve orifices must be used to prevent possible valve clogging by the 7 to 12% aluminum chlorohydrate powder normally present. Here, the vapor-tap valve, used with a mechanical break-up button, provides a fine-particled spray. The propellant content is in the 68-82% range to give good breakup and to provide an adequate reservoir for the vapor-tap.

Flammability

To devise a good aerosol product, a formulator must try to minimize the risks of flammability and possible explosivity. It is a tribute to the excellence of the aerosol packaging form that extremely flammable products can be safely dispensed, if the user follows the label directions, and if the formulator is able to make allowances for reasonably foreseeable consumer misuse. Flammability is a potential problem when large amounts of product are discharged at one time, as in some hair spray applications, painting, waterproofing, and in the total release insect fogger (TRIF) products. Flammability has also been a problem when containers are dropped on the valve stem, causing it to bend or crack in such a way that the valve jams, releasing a continuous spray. Consumers have sometimes panicked and thrown the can out the window when this happens.

The special case of the TRIF product will be described in more detail
later. The latches open on these products, allowing the entire contents of
the can, from 50 to 400 g, to be dispensed. Special low-flammability formulas
are needed to prevent harmful fireball effects if the spray is discharged too
close to pilot lights or other sources of ignition.

In the U.S., aerosol products are regulated according to type by three
federal agencies. Pesticides such as insecticides, disinfectants, herbicides,
and rodenticides are handled by the U.S. EPA. Household products such as
paints, automotive products, air fresheners, and window cleaners, are handled
by the Consumer Product Safety Commission (CPSC). Finally, all food, drug,
and cosmetic products are under the control of the Food & Drug Administration
(FDA). The EPA and CPSC require flammability labeling, according to test
method results; the FDA does not. The FDA merely states that products seen to
be too hazardous or that are inappropriately labeled will be seized and banned
from further marketing. As a result, approximately 70% of all aerosol
containers in the U.S. are marked "Flammable." Another 5%, such as many
anhydrous automotive products, are marked "Extremely Flammable." Many hair
sprays, underarm products, and other FDA-regulated aerosols are also marked
"Flammable," although this is not required.

In the U.S. and in most other countries, the standard test method for
flammability is the Flame Projection Test. Procedures and criteria vary
somewhat, but a can is normally sprayed through the top third of a candle
flame from a distance of 151 mm. If the spray ignites and carries the flame
forward another 457 mm (or further), the product is considered to be
"Flammable." The term "Extremely Flammable" is rarely used in other
countries. It relates to two tests, a flashback test and a closed cup flash
point test at approximately -28°C. For the product to be marked "Extremely
Flammable," it must fail both tests: the flash back must extend to the
actuator at any degree of valve opening, and the cup test must indicate a
flashpoint of less than -7°C.

Although there are many shortcomings of the Flame Projection Test, which
was devised in 1952, it has been adopted by many countries. A number of

techniques can be used to reduce the length of the flame in the Flame
Projection Test, so that a "Flammable" product can sometimes by "adjusted" to
a nonflammable one.   For example, hair spray marketers prefer to sell sprays
that have flame projections in the 300 to 400 mm range (thus nonflammable).
However, these products are marketed with a "Flammable" label that is, in
fact, an overspecification.

Methods for reducing flame projection include the following options:

- Reduce the delivery rate;

- Reduce particle size (smaller particles burn out more rapidly and
  move more slowly);

- Use a vapor-tap button, often with a mechanical breakup button
  (actually a way to reduce both delivery rate and particle size).

- Add a nonflammable solvent, such as 1,1,1-trichloroethane or
  methylene chloride, or a nonflammable or less flammable propellant
  to suppress flammability.

- Present the product as a lotion, foam, mousse, whip, paste, metered
  dosage (spray or foam, micro or macro) so that the test is passed
  simply because it cannot be meaningfully applied.

A relatively new concept of flammability arose in 1979 in the U.S. when
the Factory Mutual Research Corporation (owned by several insurance carriers)
was asked to look into the subject of aerosol hazards in warehouses.   Tests
showed that many aerosols exploded in fires, producing large fireballs and
intense heating effects.   Sprinkle systems need to be sized to reflect very
high fuel loading.

About 65% of the aerosol cans produced in the U.S. are anhydrous
formulations containing flammable solvents and propellants.   These require
sprinklers capable of spraying from about 3,300 liters/m^2 to 4,200 liters/m^2

(depending on the degree of water miscibility of the flammable ingredients) each minute, for control. Extremely fast response was also a requirement so that the fire could be controlled while still in an early stage.

After a $2 million fire-testing program was finally completed in 1989, the aerosol industry participated in writing new codes and in rewriting others designed to improve the safety features of warehouses, backstock storage areas, and display areas. After a lengthy development protocol, these model codes will be completed and implemented in 1991, after which it is expected they will be adopted by legislative and regulatory officials in local fire and building codes. Since the insurance companies that support the new codes are usually multinational, some effects are already being felt in Europe as well.

Pressure

Most U.S. aerosols are formulated to a pressure as low as is consistent with good operational performance across the anticipated temperature range of their use. For example, hair sprays are expected to work well between 13° - 37°C, and reasonably well just outside these limits.

Pressure limits for containers vary only modestly among countries. In the U.S., the so-called ordinary or non-specification can is permitted to hold product with pressures up to 1,067 kPa abs. (9.85 bar - gauge) at 54.4°C. It will not rupture below 1,546 kPa abs. (14.8 bar - gauge). Special cans with 14% and 28% higher pressure ratings are also available at an extra cost. They only hold about 9% of the market. Aerosols of less than 118 mL capacity are not regulated for pressure limits in the U.S. Most aerosol containers will begin to deform at about 65°C and will rupture at 75°C or higher, depending on can and product.

Materials Compatibility

The formulator's job is not complete when an acceptable product and packaging system has been developed. Test packing is always needed to establish data on weight loss rates, can and valve compatibility, organoleptic stability, etc. Hundreds of sad stories could be written about new products that were inadequately tested, and then could not be manufactured, eroded the can, demulsified, changed color or odor, were subject to microbial proliferation, grew inorganic crystals, or eventually threw down resinous precipitates in the container, swelled valves shut or partly shut, blistered can linings, became latent leakers, etc. No fewer than 36 cans per variable should be test packed and checked--some at about 25°C and some at 40°C; some upright and some inverted.

Tinplate cans do not corrode unless at least 0.008% of free water is present. Above about 0.250%, greater concentrations of water will have no additional effect on the rate of corrosion, if any. Water has little effect on aluminum cans. In fact, its virtual absence can sometimes allow anhydrous alcohol (C_2H_5OH) to attack aluminum cans to produce aluminum ethoxide $[(C_2H_5O)_3Al]$ and hydrogen (H_2) gas. Water is implicated in the well-known ability of 1,1,1-trichloroethane to sometimes attack plain and lined aluminum cans, but the mechanism is still unclear. Finally, water can facilitate development of high pH values in hair depilatory formulas and certain others, leading to aluminate (AlO_2^-) ion formation, plus hydrogen (H_2) gas. Since aluminum is amphoteric, it should only be used with formulas having a pH of less than 12.0 at 25°C, and then only when reliably lined.

If a generalized, non-pitting corrosion pattern is seen, it is best to use a lined or double-lined can. Detinning is generally a good sign, showing that the tin (not the iron) is anodic. If pitting is detected, the formula should be changed. Several options are described below:

- Remove the offending or causative ingredient if possible, such as sodium lauryl sulfate, especially if chloride ion is present.

- Add corrosion inhibitors, such as sodium nitrite, sodium benzoate, morpholine, or sodium silicates. (Do not use nitrites in conjunction with primary or secondary amines, or N-nitrosamines will very slowly form in situ. Many of these are carcinogenic.) From 0.05% to 0.20% inhibitor is generally sufficient.

- Increase the pH to about 7.6 to 8.8, if possible, by adding triethanolamine or ammonia (NH_4OH Solution).

- Remove or minimize ionizing materials, i.e., those that permit electroconductivity and thus promote galvanic corrosion reactions.

- Minimize chloride ion (especially). It is a very active corrosion promotor, even for underfilm corrosion. It is critical to minimize chloride ion when materials such as sodium lauryl sulfate (which contains it in some grades) or lauryl polyoxyethylene sulfates are present.

- Sometimes specific corrosion inhibitors are required. Sodium lauryl sarkosinate and sodium coco-B-aminopropionate surfactants are useful for sodium lauryl sulfate. Coco-diethanolamide is good for non-ionic surfactants. Virco-Pet 20 (composition proprietary, except that it is an organic phosphate), is good for dimethyl ether and water compositions.

- For some formulas, traces of moisture can be removed by using such scavengers as propylene oxide or epichlorohydrin. (Very limited evidence suggests that both may be mutagenic.) These chemicals are never recommended for cosmetics.

Many formulations that are intensely corrosive to steel cans may be conveniently packaged in lined aluminum containers. Examples are mousse products and saline solutions. The latter contain 0.9% sodium chloride (NaCl) in water under nitrogen pressure.

3. Example Non-CFC Alternative Formulations

COSMETICS, TOILETRIES, AND PERSONAL CARE PRODUCTS

Hair Sprays

In the U.S., hair spray (aerosol and pump-action) is the largest single
category of the $3,000,000,000 hair care market. Aerosol hair spray is also
the largest selling aerosol product. In 1988, about 488,000,000 units were
sold, at a retail value of about $1,150,000,000. The pump-spray alternative
has several detractions, such as finger fatigue during use, longer application
period, flexibility--and sometimes poor shape retention of the larger size
containers and occasional plugging of the meterspray valve. The formulations
must also be resistant to oxygen, since air is sucked back into the dispenser
with every actuation. The pump-action valve is rather costly, and this,
combined with a generally smaller fill volume, has necessitated a fairly high
price per unit of volume or weight. Sales are relatively small, compared with
the popular aerosol version. Both are normally anhydrous and flammable,
although there are formulation options for substantially reducing the flammability of the aerosol product.

Hair sprays are normally formulated with 1.3 to 3.0% of film-forming
ingredients, commonly called polymers or resins. These materials tack down
the hair after the product dries for a minute or two, preventing the displace-
ment of strands or curls by body motion or wind. On the other hand, plasti-
cizers are included to ensure the flexibility of the entire hair mass, so that
it can retain a healthy bounce and not feel too stiff. A feature of some
formulas is that extra stiffness can be imparted by spraying on more product,
if a more sculptured or rigid coiffure is desired.

173

The hair spray resin must have properties that include solubility in 95.5 vol% of anhydrous ethanol, a good feel on the hair, no stickiness or tackiness in moist atmospheres, lustrous (healthy) appearance, good holding power, good removability with shampoos, and sufficient flexibility to allow bounce and to resist junctural fracture.

To achieve the ideal property mix, nearly all film-forming resins are copolymers (dipolymers and terpolymers). Of the seven or so used in the U.S., perhaps the most popular is Gantrex ES-225, made by the GAF Corporation. Chemically, it is the monoethyl ester of polyvinylmaleate/maleic anhydride copolymer, and it is normally purchased as a viscous solution of 50% solids in anhydrous ethanol. For best results, the carboxylic acid moieties of the polymer must be partially neutralized by the addition of certain amine compounds. Another polymer, known as Gaffix, was introduced by the same supplier in 1989 and is said to provide not only hair fixative properties but hair conditioning as well. (The same quaternized material is recommended for hair mousses.)

The type and amount of resin and plasticizer (if needed) enables the final product to be sold as a Gentle Hold, Regular Hold, Extra Hold, Super Hold, or Ultimate Hold formulation. The differences between such products vary: a Regular Hold by one marketer may have more holding power than an Extra Hold by another. In popularity, the Extra Hold and its equivalents have a slight advantage, closely followed by the Regular. While hair sprays normally fall between the "price/value" (utility) and "luxury image" ends of the hair care market, many sell for several times the price of others. The "luxury image" products do not usually indicate their hair-holding ability, preferring to suggest that they are just right for all users.

During the 1970s, many hair sprays were extended to provide supplementary benefits by the inclusion of such minor ingredients as Vitamin E (alpha tocopherol), silicones, myristyl myristate, aloe extract, elastin, and protein hydrazolates. The products of the 1980s still use many of these special ingredients, but also claim to be "energizing," "volumizing," "revitalizing,"

"nourishing," "elasticizing," and good for "sun survival." Since hair is dead
matter, some of the claims refer to the scalp, not to the hair shaft.

The formulas in Table 6 illustrate ways of using both hydrocarbons and
dimethyl ether as propellants.

The use of water with Gantrez and Resyn copolymers in hydrocarbon-
propelled hair spray systems has been the subject of U.S. Patents held by the
American Cyanamid Company. Patents have also been issued covering the
inclusion of carbon dioxide as an additional propellant in dimethyl ether
systems. Both carbon dioxide and nitrous oxide are extraordinarily soluble in
dimethyl ether, dissolving at about 3.70% and 3.91%, respectively, for each
one bar of pressure increase at 21°C.

In the U.S., the hydrocarbon hair sprays have generally been packed in
lined tinplate or (sometimes) aluminum cans. In other countries, plain
tinplate cans are often used. The dimethyl ether formulas are usually packed
in plain tinplate cans or in aluminum cans with linings of PAM (polyamidimide)
or special epon-phenolic types.

The Precision Valve Corporation has developed effective valves for both
hydrocarbon- and dimethyl ether-based hair sprays, using their well-known 2 X
0.50 mm Aquasol® stem. Other components include a 0.50 mm MBST (Mechanical
Break-up Straight Taper) button and butyl rubber stem gasket. The very high
solvency effects of dimethyl ether require special gaskets for valves. For
valve cups, cut gaskets of Butyl U105, Butyl U133, and Chlorobutyl CLB-82 (all
by the American Gasket and Rubber Company) have performed well commercially.
The Precision Valve Corporation's Polyethylene-Sleeve gasket also gives good
performance when used with tinplate cans, as do the polyethylene and ppolypro-
pylene laminates.

Since the drying time of alcohol is 29 times as quick as that of water,
it may be surprising to know that the drying time of all four hair spray
formulas is essentially the same. This is because the formation of

TABLE 6.   HAIR SPRAY FORMULATIONS USING BOTH HYDROCARBON AND DIMETHYL
ETHER PROPELLANTS (Regular Hold)

Ingredients	Formula A	Formula B	Formula C[a]	Formula D[a]
Gantrex ES-225 (50% in Anhydrous ethanol)	4.00	4.00	----	----
Resyn 28-2930 (100%)[b]	----	----	2.50	2.50
Amino-methyl-propanol (95%)	----	0.09	0.20	0.18
N,N-Dimethyl-octadecylamine	0.29	----	----	----
Dimethyl Phthalate	0.03	----	0.03	----
D.C. Fluid #193[c]	0.02	0.02	0.04	0.06
Disodium Dodecylsulfosuccinate	----	----	0.20	0.20
Sodium Benzoate	----	----	0.08	0.08
Fragrance	0.10	0.10	0.15	0.15
Deionized Water	----	8.79	16.00	32.00
S.D. Alcohol 40-2 (Anhydrous)	67.56	61.00	44.80	28.83
Propellant A-31 or A-40[d]	28.00	26.00	----	----
Dimethyl Ether	----	----	36.00	36.00
Pressure (532 mm Vacuum Crimp, 21°C)	2.2 bar	2.5 bar	2.5 bar	3.7 bar
Delivery Rate (g/sec - 21°C)	0.50	0.54	0.60	0.65
Flame Projection (mm - 21°C)	460	425	250	225
Flash Back to Button[e] (mm - 21°C)	60	50	0	0

[a]Formulas C and D are based on information originally developed and published
by Dr. Leonidus T. Bohnenn of Aerofako, BV.

[b]Vinylacetate/crotonic acid/vinyl neodecanoate copolymer, made by National
Starch & Chemicals Corporation, U.S. (It can be replaced with Gantrez ES-225,
but some detinning may occur at 35°C or above.)

[c]A water-soluble silicone copolymer, made by the Dow-Corning Corporation, U.S.

[d]A-40 is an alternative propellant blend, consisting of 10% propane and 90%
isobutane by weight.   The pressures and other data for Formulas A and B are
based on A-31; 100% isobutane.

[e]At full delivery rate.   If the valve is throttled, the flashback of Formulas
A and B will become 152 mm; i.e., to and touching the actuator.

hydroalcoholic dimethyl ether azeotropes greatly accelerates the evaporation rate of water. (This feature is useful in dimethyl ether formulations for personal deodorants, paints, and several other aerosol products.)

Until recently, methylene chloride, and, rarely, 1,1,1-trichloroethane were included in U.S. hydrocarbon hair sprays. These solvents were removed in between 1985 and 1988 because of the alleged carcinogenicity of these substances.

Since the early 1950s, methylene chloride has been used in billions of cans of hair sprays. It increases resin solvency, decreases flammability, promotes evaporation rate, and causes the deposition of a smoother film with less junctural beading. Concentrations of up to 25% have been tolerated by both the dispenser and the consumer. In greater amounts, however, the odor and solvent effects become more significant. For instance, plastic eyeglass frames may glaze over time, and contact lenses may blush temporarily. A few individuals are sensitive to methylene chloride and may develop rashes or itching of the neck.

Hair Lusterizers

Many people, but especially those Blacks, Hispanics, and others with very curly hair, have little need for standard hair sprays, but they often use hair conditioning and lusterizing sprays that convey the sheen and look of naturally healthy hair. Some formulations for these products appear in Table 7.

Both hair sprays and hair lusterizers are sold in scented and unscented versions. The "unscented" form actually has about 0.02 to 0.03% of a nondescript floral fragrance in it to cover the slight chemical odors of the other ingredients. They are also sold for both consumer and professional end uses. The professional cans are often quite large [65 mm X 238 mm (666 mL fill)], and are generally of tinplate.

TABLE 7. HAIR CONDITIONER SPRAY FORMULATIONS

Ingredients	Formula A	Formula B	Formula C	Formula D
Isopropyl Myristate[a]	----	----	2.0	2.0
Mineral Oil; USP	----	3.0	----	----
Isodecyl Oleate	5.0	----	----	----
Volatile Silicone Fluid[b]	----	20.0	----	----
Odorless Mineral Spirits	35.0	----	20.0	20.0
PPG-12/PPG-50 Lanolins	----	----	3.0	3.0
Pluriol 9400[c]	----	----	----	0.03
Mink Oil	0.1	0.1	0.1	0.07
Fragrance	0.1	0.1	0.1	0.1
Deionized Water	----	----	----	15.0
S.D. Alcohol 40-2 Anhydrous[d]	19.8	----	44.8	24.8
Iso-Butane (A-31)	----	76.8	30.0	----
Propane/iso-butane (A-46)	40.0	----	----	----
Dimethyl Ether	----	----	----	35.0
Pressure (532 mm Vacuum Crimp) (bars, at 21°C)	3.6	2.6	2.4	2.9

[a]Cosmetic Grade. May be replaced by isopropyl palmitate.

[b]As Cyclomethicone F-251 (Dow-Corning Corporation). A blend of 25% Tetrameric Ring Compound and 75% Pentameric Ring Compound. The dimethylsilicone of 0.65 cstks. Viscosity may also be used.

[c]A propylene oxide - ethylene oxide surfactant polymer.

[d]Specially Denatured ethanol. To make, add 400 g t.Butanol and 45 g of Brucine Sulfate to 3,784 liters of ethanol.

Hair Mousse

The mousse (French word for "foam") was first introduced in an aluminum can in the U.S. in 1973 as "Balsam and Body" foam. The French-based firm of L'Oreal, S.A., which researched this product type from 1975 to 1980, required a hair setting and conditioning foam that would leave the hair softer, more manageable, easier to brush, shiny, free of frazzles, having a good handle and slip, and with good body control and able to resist fly-away situations. The product was launched in Europe in 1981 and in the U.S. and Japan in 1983. In 1988, world-wide sales were about 270,000,000 aluminum cans; about half of this number was marketed in Europe.

To achieve both hair set and conditioning characteristics, any one of three classifications of a specialty polymer must be used:

- A combination of a slightly anionic "hair spray" film former, with a compatible cationic hair conditioning polymer;

- A cationic conditioning polymer that can also function as a hair-setting agent; or

- An amphoteric hair-setting and conditioning polymer, sometimes augmented by the addition of quaternary conditioning ingredients.

One of the more popular compounds is Gafquat 755N (20% dispersion in water), which is a quaternary ammonium polymer formed from dimethyl sulfate and a copolymer of vinyl pyrrolidone and dimethylaminoethyl methacrylate. The second approach is to use a two-component system, such as a combination of GAF Corporation's Copolymer 845 (20% in water) poly(vinylpyrrolidone/dimethyl-aminoethyl methacrylate, for hair setting plus a quaternary, such as Ciba-Geigy's Bina Quat 44C: hydroxyl-cetylammonium phosphate. Supporters of the two-component system claim they can adjust the degree of set and degree of conditioning independently, to conform to perceived marketing requirements. A large number of other products are available, but the anionic and cationic moieties have to be selected for compatibility or precipitation may occur. A well-quaternized resin will exhaust substantively onto the towel-dried hair when the product is worked into it following shampooing. In a typical case, about 175 mg per 100 g of hair will exhaust from dispersions of 0.3% concentrations or higher in the product itself. This 0.175% level is all at the hair surface, and provides such properties as silkiness, shine, volume, handle, lack of fly-away, lubricity, manageability, and anti-static properties. It also avoids any sense of limpness or buildup on the treated hair.

After the hair set and conditioning agents are chosen, an emulsifier must be selected. A minimum amount should be employed, so the user can apply the mousse without needing to rinse the excess out of the treated hair. This is especially important in the case of emulsifiers, where the excess can turn the hair slightly waxy, sticky, and dull. Some common selections include oleyl diethanolamide, ethoxylated (9 mol) octylphenol, polyethylene glycol (10 mol) ether of stearyl alcohol, mixed monooleate esters of sorbitol and sorbitan anhydride with an average of 20 moles of added ethylene oxide (Polysorbate 80) and polyethylene glycol (20 mol) ether of stearyl alcohol (Brij 720 or PEG-20 Stearate). In general, the most effective are nonionic ones at levels of 0.3 to 0.7%. The emulsifier must ensure good dispersion of the propellant, good foam formation, and some initial instability of the foam when applied. It must quickly collapse when rubbed onto the wet hair. Good wet and dry combing, foam wetting, moisture retention, and emmolliency are generally conveyed by the use of these ethoxylates and propoxylates.

Like the hair sprays and lusterizers, mousse products often contain a host of specialty ingredients at levels often ranging from 0.001 to 0.100%. These include aloe vera extract, jojoba oil, chamomile extract, protein derivatives, elastin, allantoin, other quaternaries, birch (tree) extract, marigold (flower) extract, walnut leaves (tree) extract, and various sunscreening agents. They may or may not convey any real benefit, depending on the concentration used in the formula. Some mousse products also contain colorants or dyes, of which perhaps the most common is FD&C Yellow #7 (in the U.S.), used at 0.002 to 0.008%. All mousses are perfumed.

Bacterial proliferation can occur in some mousse formulas, so they are often protected with 0.10% methyl p.hydroxybenzoate and 0.05% n.propyl p.hydroxybenzoate. Other, more powerful and broader spectrum preservatives are now being favored, such as Kathon CG (Rohm & Haas Company) and Dowicil 200 (Dow Chemical Company). In general, the finished mousse concentrate or aerosol should be able to pass a microbial Total Plate Count test with a reading of "less than 10 organisms per mL." It is also recommended that the tank, hoses, pumps, filters, and filler bowls be sanitized and that the deionized water used in batchmaking be first heated to 70°C to kill pseudomonads and most other microorganisms.

The usual propellant for mousse products ia A-46 (15 weight % propane, in 85 weight % isobutane), which develops mousse pressures in the area of 4.0 bar at 21.2°C. The usual amount is 6.5 to 7.5%, but some products use as much as 15%. For some specialty products, such as mousse used on babies, absolutely no evidence of flammability can be tolerated. With the straight hydrocarbon formulas, touching a lit match or lighter to the surface of the foam will produce a momentary ignition. This can be eliminated by using a propellant blend that includes 30% or more HFC-152a. The A-46 and HFC-152a can be purchased as a blend, premixed by those fillers who have blending stations, or added consecutively using two separate gassing machines. Because the HFC-152a is only present in concentrations of 2% or so, the cost penalty is relatively low.

The general considerations involved in formulating a mousse hair set and conditioning product have been described, and some illustrative formulations will now be presented. Table 8 describes four formulations fully, giving first the U.S. Cosmetic Ingredients Dictionary (CTFA-CID) terminology, followed by the chemical name, brandname(s), and source. Table 9 describes four additional formulations in a format of decreasing order of ingredient concentration (except that ingredients whose concentrations are less than 1% may be placed in any order), in accordance with U.S. Food and Drug Administration (FDA) regulations. These regulations require ingredients of food, drugs,

TABLE 8.   MOUSSE HAIR SET AND CONDITIONING PRODUCT FORMULATIONS

Formula A		
	% (w/w)	
	Soft Set	Firm Set
Polyquaternium 4 Copolymer of hydroxy-methylcellulose and diallyl-dimethyl ammonium chloride	0.60	1.00
Celquat L-200 (100% A.I.) by National Starch & Chemical Corporation		
Deionized Water	75.85	75.25
Dimethacone Dimethyl silicone derivative emulsion	0.15	0.20
DC Silicone Emulsion by Dow-Corning Corp.		
Tallow Trimonium chloride (and) isopropanol stearyl/palmityl trimethyl ammonium chlorides; 75%, in isopropanol	0.10	0.15
Arquad T-50 (75% Active Ingredient) by the Industrial Chemicals Division of Armak Corp.		
Octoxynol 9 Ethoxylated (9) n.Octylphenol	0.15	0.30
Triton X-100, by the Rohm & Haas Corp.		
Emulsifying Wax NF Fatty alcohol derivative - Self emulsifying	0.15	0.10
Polawax A-310, by Croda, Inc.		
Polawax A-310, (100% A.I.) by Croda, Inc.		
Ethanol SD40 Ethanol (Denatured #40; 100%)	14.90	14.88
S.D. Alcohol 40; anhydrous, by U.S. Industrial Chemicals Division		
Perfume Oil (Floral)	0.10	0.12

Continued

TABLE 8.   (Continued)

Formula A

Ingredients	% (w/w)	
Propellant A-46 15 wt. % propane & 85 wt.% isobutane	8.00	8.00
A-46 Propellant by Phillips Petroleum Co.; Specialty Products Division		

Formula B

Ingredients	% (w/w)
Quaternium 11 Poly (vinylpyrrolidone/dimethylaminoethyl methacrylate)	7.00
Copolymer 845 (20% Solids in Water) by GAF Corp.	
Polyquaternium 16 Hydroxyethyl cetyldimonium phosphate (100% A.I.)	3.50
Bina-Quat 44C (100% A.I.) by Ciba-Geigy Corp.	
Cocoamid DEA Coconut acids diethanolamine condensate (1:1) (Free of soap and amide esters) (Superamide) Standamid SD (100% A.I.) (Henkel Chemical Co.)	1.00
Tallow Alkomium Chloride Dimethyl benzyl tallow ammonium chloride	0.50
Incroquat S85 or SDQ-25 (Croda, Inc.)	
Deionized Water	77.65
Methylparaben Methyl p.hydroxybenzoate	0.08
Nipagin M (Nipa Laboratories, Ltd.)	
Perfume Oil (Floral)	0.26
FD&C or D&C (Color)	0.01
Propellant A-46 16 wt % propane & 84 wt % isobutane	10.00
A-46 Propellant by Phillips Petroleum Co.; Specialty Products Division	

Continued

TABLE 8.   (Continued)

Formula C	
Ingredients	% (w/w)

Polyquaternium 11   1.32
    A quaternary ammonium polymer formed by the
    action of dimethyl sulfate on a copolymer of
    vinylpyrrolidone and dimethyl amino ethyl methacrylate.

    Gafquat 734, (50% A.I. in Ethanol), by the GAF Corp.

Polyquaterium 4   1.00
    A copolymer of hydroxymethylcellulose and diallyl-
    dimethyl ammonium chloride.

    Celquat H60 (100% A.I.) by National Starch and Chemical Corp.

Silicone   0.15
    Silicone polymer, end-blocked with aminofunctional
    groups

    Cationic Emulsion 929 Dow-Corning Corporation

Oleamidpropyl Dimethylamine Hydrolysed Animal Protein   0.20
    Oleylamidopropyl diethylamine hydrolysed animal protein.

    Lexein CP-125, by the Inolex Corp.

Potassium Coco-hydrolysed Animal Protein   0.14
    Animal protein hydrolysed in boiling potassium cocoate
    soap solution.

    Lexein S620, by the Inolex Corp.

Aloe vera   0.05
    Aloe vera

    Aloe Vera: Pure Extract (90% A.I. Powder),
    Terry Chemical Company

PEG 150   0.26
    Hydro-(ethyleneoxide 150) alcohol

    Carbowax 8000, by Union Carbide Corp. or Polyethylene
    Glycol 6000 by Dow Chemical Company

Quatermium 52   0.20
    Dibutyl sebacate

    Dehyquart SP, by Henkel Chemical Company

Continued

TABLE 8.   (Continued)

Formula C	
Ingredients	% (w/w)

Ethanol SD40
    Ethanol (specially denatured #40; 100%)
 	
    S.D. Alcohol 40; Anhydrous, by Shell Chemical Co.
	3.00

Polysorbate 20
    Mainly the monolaurate ester of sorbitol and
    sorbitol anhydrides, condensed with 20 moles of
    ethylene oxide.
 	
    Tween 20 by ICI Americas, Inc., or Nikkol TL10
    or TL10-EX by the Nikko Chemical Company
	0.05

Fragrance (Floral)	0.209
FD&C or D&C (Color)	0.001
Deionized Water	85.42

Propellant BIP-55		8.00
Ethane	0.290 w.%	
Propane	30.728	
Iso-butane	26.509	
n.Butane	39.759	
Pentanes	2.700	
Hexanes	0.010	
Unsaturated Hydrocarbons	0.001 maximum	
Sulfur compounds	0.0005 maximum	
Water	0.0025 maximum	

Propellant IBP-55 by Phillips Petroleum Company,
Specialty Products Division.

Formula D	
Ingredients	% (w/w)

Polyquaternium 11
    Quaternary ammonium polymer of dimethyl sulfate
    and the copolymer of vinylpyrrolidone and
    dimethylaminoethyl methacrylate.
 	
    Qafquat 755N (20% in water), by GAF Corp.
	5.00

Continued

TABLE 8.   (Continued)

Formula D	
Ingredients	**% (w/w)**
PVP	1.00
Polyvinylpyrrolidone (Mol. wt. = 30,000)	
PVP (K-30) (GAF Corp.)	
Carbomer 941	15.00
Polymer of acrylic acid, cross-linked with a polyfunctional agent	
Carbopol 941 (Use as 2.0% in Water) B.F. Goodrich & Co.	
Ammonia	0.28
Ammonium Hydroxide (29% in Water)	
Steardimonium Hydrolysed Animal Protein - Purified.	0.28
Stearyl dimethyl ammonium modified hydrolysed protein	
Croquat SP (Croda, Inc.)	
Nonoxynol-20	0.28
Ethoxylated (20) n.nonylphenol	
Igepal CO-850 (GAF Corp)	
Steareth-2	0.28
Polyethylene glycol (2) ether of stearyl alcohol $CH_3(CH_2)_{16}CH_2(OCH_2CH_2)_2OH$ Brij 72 (ICI Americas, Inc.)	
Polysorbate 20	0.50
Mainly the monolaurate ester of sorbitol and sorbitol anhydride, condensed with an average of 20 moles of ethylene oxide.	
Nikkol TL10 or TL10-EX (Nikko Chemical Co.)	
Methylchloroisolthiazolinone and Methylisothiazolinone	
Iathon CG (Rohm & Haas Company)	
Fragrance	0.24
Deionized Water	68.00

Continued

TABLE 8.   (Continued)

Formula D	
Ingredients	% (w/w)
Hydrofluorocarbon 152A	6.40
1,1-Difluorethane	
Dymel-152 (E.I. DuPont de Nemours & Co., Inc.)	
Genetron-152a (Allied-Signal Corporation)	
Isobutane	2.74
Isobutane A-31 (Phillips Petroleum Co.)	
Aeron A-31 (Diversified Chemicals and Propellants Co.)	

TABLE 9.   INGREDIENT LISTINGS OF OTHER MOUSSE HAIR SETS
            AND CONDITIONERS[a]

---

1. XYZ Co. Professional Designer Mousse

   So Airy-light   Extra Firm   Alcohol Free   With Sunscreen

                Water
                Hydrofluorocarbon 152A
                Isobutane
                *Polyquaternium-11
                *DEA-Methoxycinnamate
                 Polyquaternium-4
                 Dimethacone Copolyol
                 Fragrance
                *Isosteareth-10
                 Sodium Cocoyl Isethionate
                 Methyl Paraben
                *Lauramide DEA
                *DMDM Hydantoin

2. DO-GLO XYZ Co. Alcohol-Free Styling Mousses

                A 230-gram fill in aluminum can

                Ultimate Hold - for all Hair

                Water
                Isobutane
                PVP/Dimethylaminoethyl methacrylate copolymer
                Polyquatermium 4
                *Diphenyl-dimethicone
                *Lauramine Oxide
                 DMDM Hydantoin
                 Fragrance
                *Quaternium 18
                 Butane
                *Ammonium Laureth Sulfate
                *Disodium Ethylenediamine Tetraacetate (EDTA)
                 Citric Acid

Continued

TABLE 9.   (Continued)

3.   DO-GLO XYZ Co.   Alcohol-Free Styling Mousses

A 230-gram fill in aluminum can

Extra Body - for Fine Hair

Water
Isobutane
Polyquatermium 4
Propane
*Lauramine Oxide
Propylene Glycol
*Octyl Salicylate
Panthenol
*Silk Amino acids
*Keratin Amino acids
*Hydrolysed Animal Keratin
Butane
Citric Acid
DMDM Hydantoin
Fragrance
*Disodium Ethylenediamine Tetraacetate (EDTA)

4.   DO-GLO XYZ Co.   Alcohol-Free Styling Mousses

A 230-gram fill in aluminum can

Moisture Rich - for Dry or Damaged Hair

Water
Propane
Isobutane
*Acetamide Monoethylamide (MEA)
PVP/Dimethylaminoethylmethacrylate Copolymer
Butane
Cocamide Diethanolamide (Superamide)
Panthenol
(N-Pantothenylamindoethyl) disulfide
*Glycereth-26
*PEG-150 Distearate
Sodium Lactate
*Sodium PCA
Collagen

TABLE 9.   (Continued)

˙All ingredients are listed in decreasing percentages, except for those
 present in concentrations of less than 1.0 percent.

*These ingredient designations are identified chemically in the following way:
     Acetamide MEA: Acetamide Monoethylamide Lipamide MEAA (Lipo Chemicals,
                         Inc.)
     Ammonium Laureth Sulfate: Ammonium salt of ethoxylated (1-4) lauryl
                         Sulfate Carsonol ALES-4 (Lonza Chemical Corp.)
     DEA-Methoxycinnamate: Diethylaminomethoxycinnamate (Sun Screen Agent)
     Diphenyl-trimethicone: Silicone 556 Fluid (Dow-Corning Corporation)
     Disodium EDTA: Disodium Ethylenediaminetetraacetate
     DMDM Hydantoin: 1,3-Dimethylol-5,5-Dimethyl Hydratoin
     Glycereth-26: Polyethylene Glycol (26) Glyceryl Ether Ethosperse G-26
                       (Glyco Chemical Company)
     Hydrolysed Animal Keratin: Keratin, hydrolysed
     Isostreareth-10:   Polyethylene Glycol (10) Ether of Isostearyl Alcohol
     Keratin: Keratin Amino Acids Kerapro (Hormel Company)
     Lauramide DEA: Lauric Acid - Di-ethanolamide (1:1) Condensate Super-
                         amide
     Lauramine Oxide: n.Lauryl-dimethylamine oxide
     Octyl Salicylate: 2-Ethylhexyl Salicylate $C_{15}H_{22}O_3$ (Sunarome WMO-Felton)
     PEG-150 Distearate: Polyethylene Glycol (150 mol) Distearate Lipopeg
                         6000-DS (Lipo Chemicals, Inc.)
     Polyquaternium 11: A quaternary ammonium polymer formed from dimethyl
                         sulfate and a copolymer of vinylpyrrolidone and
                         dimethyl amino ethyl methacrylate.
     Quaternium 18: De(hydrogenated tallow) Dimethylammonium Chloride
                         Adogen 442 (Sherex Corporation)
     Silk Amino Acids: Amino Acid Blend, derived from Silk Protein (Croda,
                         Inc.)
     Sodium PCA: Nalidone (UCIB) U.S. Distributor: S.S.T. Corporation
                    (Clifton, NJ)

and cosmetics to be listed on the product label in this fashion, unless they are for professional or institutional use.

Containers for Hair Setting and Conditioning Mousses.

Mousse formulas are normally finished to a pH value of approximately 5.5 to 6.5 at 25°C. They are either aqueous or hydroalcoholic and contain surfactant wetting agents and a high concentration of chloride ion, a well-known corrosion promoting agent. It is not surprising that they are corrosive to steel and tinplated cans and may even attack aluminum cans unless they are well lined. Many mousse formulations contain corrosion inhibitors, such as sodium benzoate, coco-diethanolamide, and amino groups to provide additional shelf-life stability.

During 1984 and 1985, a large amount of work went into developing well-lined tinplate cans for mousse formulas to take advantage of the much lower prices of tinplate. In 1984, one major marketer introduced a low-priced mousse in a 45-mm diameter tinplate (necked-in) can. Special techniques were used to make the formula less aggressive, and the can was heavily double-lined with an Organosol hydridized vinyl coating system. The product is still doing well on the market. The only other mousse products in tinplate cans are lines of large, salon-type mousse products sold to a number of marketers by one formulating house. The cans are 52 mm in diameter by approximately 190 mm in length. They have necked-in construction. They are made only by the Continental Can Unit of the United States Can Company and, uniquely, use a third body lining (after welding and flanging) as a repair coat, to cover up any abrasions or scratches made during manufacture. These cans contain 298 g of product. A developmental can of polypropylene-laminated tinplate or tin-free steel is performing well with mousse products after a year of storage.

Aluminum cans for mousse applications are typically 38, 45, or 52 mm in diameter and are up to 165 mm long. All are of one-piece construction and are heavily single or double lined. The usual epon-phenolic linings for aluminum cans are sometimes inadequate for mousse products and have generally been replaced with linings made of pigmented epoxy-phenolics, PAM clear polyamidimide, and the popular Micoflex L6X392 beige-pigmented vinyl Organosol. In some cases, two separate linings of the same material are applied.

Nearly all valves for mousse hair care products use aluminum mounting cups coated on the outside and lined on the inside surfaces. Numerous problems have arisen from trying to make a good seal, especially in the case of the larger-diameter 52- and 66-mm cans. Bonded polyethylene sleeve gaskets (an exclusive development of the Precision Valve Corporation) are satisfactory if the new Ring Seal mounting cup contour is used, as are the full-coverage polyethylene laminate gaskets--again, if a special contour valve cup is used. These gaskets require a special mousse-resistant adhesive if continuing attachment to the inner surface of the valve cup is a marketer requirement. Otherwise, the laminate will separate eventually and droop slightly. Some technical experts are concerned about the possible corrosivity of concentration cells that can be created between the laminate and unprotected aluminum cup. Any liquids that may accumulate there must enter by permeating the polyethylene; therefore, unknown compositions and concentrations may form. Also, the adhesive is generally an excellent barrier material, but this advantage is lost if delamination occurs.

Most hair care mousse products are designed to be vertical acting. Full coverage of the mounting cup is an aesthetic benefit. The "pad-and-smokestack" type is popular, as is the tilt-action, simple "smokestack" design. Foam spouts are either white or the color of the base coat of the can. The protective cover will either fit over the spout or valve cup outside diameter, or be designed to have the same diameter as the can, for a cylindrical look.

Many U.S. marketers are testing the Fibrenyle Ltd. "Petasol" or PET (polyethylene terephthalate) plastic bottles for mousse applications, with good results. One problem that must be resolved is that the U.S. Department of Transportation (DOT) will not permit interstate shipments of non-metallic aerosol containers if their capacity exceeds 118.3 mL because of a regulation that dates back to 1951, when the only non-metallic aerosol containers were those made of glass. Special exemptions based on impressive arrays of test data, including drop test results, have now been requested. Meanwhile, several new bottle shapes are being developed.

Other Mousse Products

The original hair setting and conditioning mousse products of 1981 and later have made it possible for marketers to successfully introduce an impressive number of related foam products. Several still involve various aspects of hair care, such as hair sheens or lusterizers, hair depilatories, dandruff control foams (using zinc pyrithrone or an alternative dandricide), hair coloring mousses, and mousse products to help control ear itch and "jock itch," a trichophytal fungal mycosis of the pubic area related to the well-known "Athlete's Foot" problem. Other hair care products include baby oil formulas, baby shampoos, curl activators, and products that promise a hair thickening effect. A hair restorative material is now being sold in mousse form.

The remaining second-generation mousse products are generally for skin care. They include make-up items, baby oil formulas (again), sunscreens, facial cleansing foams, hand creams, skin smoothing (20% talc) products, cationic skin emollients, dewrinkler formulas, etc. Mousse products can be used to contain and deposit large amounts of specialty oils on the skin, such as a product that contains 25% jojoba oil (Wickenol 139, Dow-Corning Corp.), which claims to give the skin a healthy sheen and superior lubricity.

The specialty "mousse gels" consist of lines of products that deliver as clear or translucent gels but spring into mousse foams on contact with the hand. Table 10 presents several formulations illustrating these newer products.

The mousse packaging system is an ideal vehicle for dispensing sunscreens and suntan lotions as foams that quickly break up on mechanical shearing to produce exceptionally even matte finish results. The first to introduce this product was Schering-Plough, Inc. as one of several package forms for Copper-tone. More recently, in 1984, the Golden Sun Company introduced their Sun Goddess Body Mousse Protective Sun Tan products in a 170-g filling weight.

TABLE 10.   SPECIALTY MOUSSE FORMULATIONS

Formula E

Hair Sheen and Conditioner

Ingredients	% (w/w)
Stearimidopropyl Cetaryl Cimonium Tosylate (and) Propylene Glycol. (Ceraphyl 85, by Van Dyk & Co., Inc.)	0.50
Quaternium 26 (Ceraphyl 65, by Van Dyk & Co., Inc.)	0.75
C9-11 Isoparaffins (99+% C_9-C_{11} Branched Paraffinics) Isopar K, by Exxon, Inc.	20.00
Isodecyl Oleate	4.00
C12-15 Alcohols Benzoate	1.00
Fragrance	0.25
Deionized Water	68.50
Isobutane A-31 (Aeropres Corp.) (Propellant A-46)	4.25
Propane A-108 (Aeropres Corp.) (Propellant A-46)	0.75

Note: The foam may be destabilized by adding 0.5% of a volatile silicone such as CTFA Cyclomethicone, alcohol, or by replacing part or all of the hydrocarbon propellant (A-46) with hydrofluorocarbon 152 (CH_3 - CHF_2). Stability can be increased by adding cetyl alcohol.

Formula F

Baby Shampoo and Conditioner

Ingredients	% (w/w)
Sodium Laureth (3) Sulfate [Sodium Lauryl (3 ETO) Sulfate]	40.00
Cocoamidopropyl Betaine (Cocoamidopropyl dimethylglycine) Aerosol 30 by American Cyanamid Company, or Velvetex BA-35 by the Henkel Chemical Company.	16.00
Benzyl Alcohol	0.25

Continued

TABLE 10.   (Continued)

Formula F (Continued)

Ingredients	% (w/w)
Methyl p.Hydroxybenzoate	0.25
PEG-150 [H-(OCH$_2$ -CH$_2$)$_{150}$OH] Carbowax 8000 by Dow Chemical Company.	0.25
Fragrance	0.25
Deionized Water	37.00
Isobutane A-31 (Aeropres Corporation) (Propellant A-46)	5.10
Propane A-108 (Aeropres Corporation) (Propellant A-46)	0.90

Note: The first two ingredients may be replaced with 34.00% Disodium Oleamido PEG-2 Sulfosuccinate, 20.00% Sodium Laureth (3) Sulfate, and 2.00% Quaternium 22 substantive conditioner and humectant. The last item is available as Ceraphyl 60 from Van Dyk & Company, Inc.

Formula G

Easy-Spreading, Hair-Fixative Mousse

Ingredients	% (w/w)
Ethyl Ester of PVM/MA Copolymer (Gantrez ES-225, by the GAF Corporation; 50% A.I. in Ethanol)	5.00
Dimethicone Polyol (Surfactant 193 Fluid, by Dow-Corning Corporation)	0.50
Amino methyl propanol (95%) min.) AMP-95, by the IMC Corp.	0.20
Fragrance	0.20
Ethanol (Anhydrous Basis) S.D. Alcohol 40 (200°), by Publicker Industries, Inc.	10.00
Deionized Water	75.10
Isobutane A-31 (Technical Petroleum Company)	7.65

Continued

TABLE 10.   (Continued)

<hr>

Formula G (Continued)

Ingredients	% (w/w)
Propane A-108 (Technical Petroleum Company)	1.35

Note: The water-soluble 193 Surfactant acts to plasticize the copolymer
and to enhance the spreadability of the resin on the hair. It also
enhances foam building and foam stability. The possible need for a
preservative should be investigated, although most of these formulas
do not need one.

<hr>

Formula H

Alcohol-Free Mousse for Damaged Hair

Ingredients	% (w/w)
Polyquaternium 11 (Gafquat 755N by GAF Corp.)	7.50
Blend of Trimethylsilylamodimeticone, Octoxynol 40, Isolaureth-6 and a glycol. (Dow-Corning Q2-7224; Dow-Corning Corporation)	1.00
Oleth-20 (PEG 20 Ether of Lauryl Alcohol) (Brij 98, by ICI Americas, Inc.)	0.50
Fragrance	0.20
Deionized Water	81.80
Isobutane A-31 (Technical Petroleum Company) (Propellant A-46)	7.65
Propane A-108 (Technical Petroleum Company) (Propellant A-46)	1.35

Note: The Dow-Corning Q2-7224 conditioning agent provides improved wet and
dry combing and imparts a good handle or feel to the hair. The
agent is particularly effective on damaged hair.

Combinations of about 3.0% Polyquaternium 11, 0.3% Polyquaternium
10, 0.3% Steareth 10 (Brij 76) and 0.15% PEG-2 Oleammonium Chloride
(Ethoquad 0/12 - Armak) form a good base for alternative formulas.
In some cases, up to 6.0% ethanol may be added.

<hr>

Continued

TABLE 10.   (Continued)

Formula I

Mousse Curl Activator

Ingredients	% (w/w)
Propylene Glycol	32.00
Isodecyl Oleate or Myristyl Myristate	4.00
C12-C15 Alcohols Lactate Ceraphyl 41 by Van Dyk & Co., Inc.	6.00
Glycerine	5.00
Quaternium-26 (Hydroxyethyl) Dimethyl (3-Mink animal oil amidopropyl) Chlorides. Ceraphyl 65 by Van Dyk & Co., Inc.	2.00
Quaternium-22 3-(D-Gluconoyl-amino)-N-(2-hydroxyethyl)-N,N-Dimethyl-1-propanaminium Chloride. Ceraphyl 60 by Van Dyk & Co., Inc.	1.50
Dimethacone Polyol Silicone Fluid L-720 by Union Carbide Corp.	1.50
PEG-40 Stearate Polyethylene Glycol (40 mol) Mono-ester of Stearic Acid	1.50
Cetyl Alcohol	0.50
Fragrance	0.25
Methyl p.Hydroxybenzoate Nipagin M by Nipa Laboratories, Ltd.	0.25
Deionized Water	40.50
Isobutane A-31 Phillips Petroleum Company (Propellant A-46)	0.75
Propane A-108 Phillips Petroleum Company (Propellant A-46)	4.25

Note: The propylene glycol and glycerine are hair curl activators, while
        the Isodecyl Oleate and C12-C15 Alcohols Lactate are glossing
        agents. The Quaternium 22 is an optional agent, used to increase
        humectancy and conditioning.

Four products were presented, with SPFs (Sun Protection Factors) of 2, 4, 6, and 12. For example, a product with an SPF of 12 will enable the user to remain in the sun twelve times as long as the standard period required to develop slight redness, while developing the same degree of coloration.

Since 1984, a number of relatively low-sales-volume sun screen mousse formulas have been launched. Some are identified as "sun screens;" others are labeled "sun and sport styling mousses" or other, less definitive names. Table 11 lists the ingredients of two strengths of typical sunscreen mousses.

Sun protection formulations with SPF values of 15 to 20 (the practical maximum) are often called sun-blocks (see Formula J in Table 11). They require combinations of sun-screen agents of the oil- and water-soluble type, to get the best distribution on the skin. In some cases, rigorous pH control, using buffering agents, is required. Concentrations of from 5.5 to 10.0% are required, depending on the efficiency of the screening agents, type of mousse emulsion, distribution on the skin, repeat applications, skin condition and moisture content, individual sensitivity, skin color, season, time of day, elevation, geographic latitude, type of activities, perspiration rate, product application thickness, etc.

For most people, a lesser degree of protection is acceptable. For those spending two or three hours in the sun at one time, the use of products with an SPF of about 4 to 6 is satisfactory. These products also permit the development of a tan, which is often socially important to those people or races having light-colored skin pigmentation.

Formula K, shown in Table 11, provides this intermediate degree of sun protection, based on the use of iso-amyl-p-methoxy-cinnamate. This water-insoluble material provides SPFs of 4 (at 2.8%), 6 (at 3.6%), 8 (at 4.5%), and 12 (at 7.5%). Thus, the formula can be adjusted to give whatever degree of solar protection is desired.

TABLE 11.   SUNSCREEN MOUSSE FORMULATIONS

Formula J	
Sunscreen Mousse (SPF About 15)	
Ingredients	**% (w/w)**
Isodecyl Oleate or Myristyl Myristate Isodecyl Oleate is available from Van Dyk & Co., Inc. as Ceraphyl 140-A.	6.00

Octyl Dimethyl PABA (formerly Padamate 2)
    Ester of 2-Ethylhexyl Alcohol & Dimethyl p.aminobenzoic
Acid
 	
    Escalol 507 by Van Dyk & Co., Inc.
	6.85

Benzophenone-3
    2-Hydroxy-4-methoxybenzophenone
 	
    Uvinol M40 by BASF-Wyandotte Chemical Company
	3.20

Stearic Acid Octadeconoic Acid	10.00
Cetyl Alcohol Hexadecyl Alcohol	0.45
Deionized Water	44.95
Hydroxypropyl Methylcellulose Methocel F, by Dow Chemical Company, or Viscontran MHPC by Henkel Chemical Company	0.55
Propylene Glycol	2.50
Triethanolamine 99% Triethanolamine by Union Carbide Corp.	1.15

Ethanol
    Alcohol (Anhydrous Basis; Specially Denatured)
 	
    S.D. Alcohol 40 by U.S. Industrial Chemicals Division
	18.00

Continued

TABLE 11.   (Continued)

<hr>

Formula J (Continued)

Ingredients	% (w/w)

TEA Coco-Hydrolysed Animal Protein (and) Sorbitol
    Triethanolamine Salt of the condensation product of
    coconut acids and hydrolysed animal protein
	1.00

Maypon 4CT by Stepan Chemical Company	
Methyl p.Hydroxybenzoic Acid	0.15
Perfume	0.20
Isobutane	4.25
Propane	0.75

Formula K

Sunscreen Mousse (SPF About 6)

Ingredients	% (w/w)

Steareth-10
    PEG Ether of Stearyl Alcohol; $CH_3(CH_2)_{16}CH_2$-
    $(OCH_2CH_2)_{10}OH$
	0.80

Brij 76, by ICI Americas, Inc.	
PEG 150 Distearate Polyethylene Glycol (150) Diester of Stearic Acid	0.60
Kessco X-211 by Armak Chemical Company, or Witconol L32-45 by Witco Organics Division.	

Sodium Hyaluronate
    Sodium hyaluronate - high-molecular-weight polymer
    from animal protein (90% powder) by Tri-K Industries,
    Inc.   (Reseller for Canadian Packers, Ltd.)
	0.10

DMDM Hydantoin 1,3-Dimethylol-5,5-Dimethyl Hydantoin	1.10
Dantoin DMDMH-55 (or Glydant) by Glyco Products, Inc.	

Continued

TABLE 11.   (Continued)

Formula K (Continued)

Ingredients	% (w/w)

Quaternium 52
  Dibutyl Sebacate
 	
  Dehyquart SP, by Henkel Chemical Co.
	8.50

Isoamyl Methoxycinnamate (CTFA) (FDA is pending)
  Isoamyl-p-methoxycinnamate (98% min.)
 	
  Neo-Heliopan E1000 by Haarmann & Reimer Japan K.K.
	3.60

Cyclomethicone
  Cyclic dimethylpoly (3 - 4) siloxane
 	
  Silicone #344 Fluid by Dow-Corning Corporation
	2.50

Dimethacone Copolyol (Water Soluble)
  Dimethylsiloxane, end-blocked with surfactant groups
 	
  Silicone #193 Surfactant by Dow-Corning Corporation
	0.25

Polysorbate 80
  Mixed oleate esters of sorbitol and sorbitol anhydride;
  mostly the monoesters, with 20 moles of ethylene oxide
 	
  Tween 80 by ICI Americas, Inc.
	0.15

Polysorbate 20
  Mixed laurate esters of sorbitol and sorbitol anhydride;
  mainly the monoesters, condensed with about 30 moles of
  ethylene oxide.
 	
  Nikko TL10 or TL-10EX by Nikko Chemicals, Ltd. or
  Tween 20 by ICI Americas, Inc.
	0.40

Olealkonium Chloride
  Oleyl-dimethyl-benzyl-ammonium Chloride
 	
  Ammonyx KP by the Onyx Unit of Stepan Chemical Co.
	0.10

Continued

TABLE 11.   (Continued)

Formula K (Continued)

Ingredients	% (w/w)
Nonoxynol-20 Ethoxylated (20)-p-n.nonylphenol	0.10
Igepal CO-850 by GAF Corporation, or Tergitol NPX by Union Carbide Corporation	
Aloe Vera Aloe Vera Extract - 100% Pure (90% A.I. Powder)	0.10
Aloe Vera 100% by the Terry Corporation	
Perfume	0.30
Methylchlorisothiazolinone, and Methylisothiazolinone Kathon CG by Rohm & Haas Company	0.10
Hydrofluorocarbon 152 1,1-Difluoroethane	4.00
Dymel 152 by E.I. DuPont de Nemours & Company, Inc.	
Isobutane Isobutane A-31 by Phillips Petroleum Company	2.00

The mousse presentation can also be used for baby care products. They avoid the spillage and application problems of other formulations. Table 12 lists the ingredients of these formulations.

Other commercial mousse products include a facial cleansing preparation based on very mild surfactants such as disodium cocoamido MIPA sulfosuccinate and sodium laureth sulfate, including Quaternium-22 to provide conditioning and emolliency. A facial makeup mousse uses a triethanolamine stearate and PEG-20 stearate emulsifier combination to spread a combination of pigments and emollients on the face to give an elegant matte finish. Hand creams are also available, again based on triethanolamine stearate and PEG fatty acid conden- sates as the emulsifier. Glycerin (5%) is included as a humectant. More recently, a cationic skin mousse has appeared that includes mink (animal) amidopropyl dimethylamine to provide a unique lubricity and skin feel. Several more interesting products are under intensive development.

The vaginal contraceptive foam is a product in the drug category that depends on foam stability and density. The preferred active ingredient is Nonoxynol 9, or nonyl-phenoxypolyoxyethylene ethanol. The formula for a mousse product of this type has been published, and a variation is presented as Formula N in Table 13.

Commercial formulas in the U.S. range from 8-12.5% of Nonoxynol 9, and all the aerosols are pressurized with about 8% of a blend of CFC-12 and CFC- 114. An Amended New Drug Application (to the FDA) will be required before the propellant can be changed to hydrocarbon or other types. This process takes the FDA about 3 to 5 years to complete, since the entire NDA file must be reviewed whenever a change is made.

A mousse product is also available for the treatment of mastitis infec- tions in the udder of cows. An example is presented in Table 14.

A similar formula, based on the use of Procaine Penicillin G, represents an anhydrous mousse system. Both formulas are designed for injection into the udder via the sphincter canals. CFC-12 and CFC-114 are used for the products

TABLE 12.   BABY CARE MOUSSE FORMULATIONS

Formula L	
Baby Oil Mousse	
Ingredients	**% (w/w)**
Deionized Water	23.00
Mineral Oil - Medium Viscosity	30.00
Cetaryl Alcohol (and) Cetareth-20 C16-C18 Alcohols and C16-C18 Alcohol PEG 20 Ethers	10.00
Macol 124 by the Mazer Chemical Company	
Isodecyl Oleate and/or Myristyl Myristate Ceraphyl 140-A and/or Ceraphyl 424 by Van Dyk & Co., Inc.	10.00
Cyclomethicone Cyclic dimethylpoly (3-4) Siloxane	3.35
Silicone #344 Fluid by Dow-Corning Corporation	
Ethanol S.D. Alcohol 40 (Anhydrous)	12.00
S.D. Alcohol 40 (200°) by U.S. Industrial Chemical Div.	
Aloe Vera Aloe Vera Extract - 100% Pure (90% A.I. Powder)	0.25
Aloe Vera 100% by the Terry Corporation	
Perfume Perfume selected for mildness	0.25
Methyl p.Hydroxybenzoate Nipagin M by Nipa Laboratories, Ltd.	0.15
Hydrofluorocarbon 152 1,1-Difluoroethane	8.00
Dymel 152 by E.I. DuPont de Nemours & Co., Inc.	
Isobutane Isobutane A-31 by Phillips Petroleum Company	3.00

Continued

TABLE 12.   (Continued)

Formula M	
Baby Powder Mousse	
Ingredients	**% (w/w)**
Deionized Water	35.70

Silica (Powder)
    Hi-Sil 233 Fumed Silica Anticaking Agent by PPG
    Industries, Inc.
	0.30

Quaternium-26
    (Hydroxyethyl)Dimethyl(3-Mink Animal Oil Amidopropyl)
    Chlorides
	2.00

Ceraphyl 65 by Van Dyk and Company, Inc.	
PPG Laneth 50 Polyoxyethylene (50) polyoxypropylene (12) Lanolin Ether	0.50
Solulan by Amerchol Products Unit	
Cetyl Alcohol	0.40
Talc Talcum Powder - Impalpable	18.10
Altalc 200	
Ethanol S.D. Alcohol 40 (Anhydrous)	32.00
S.D. Alcohol 40 (200°) by Shell Chemical Company	
Perfume	0.20
Hydrofluorocarbon 152 1,1-Difluoroethane	8.00
Dymel 152 by E.I. DuPont de Nemours & Co., Inc.	
Isobutane Isobutane A-31 by Phillips Petroleum Company	2.80

TABLE 13.   VAGINAL CONTRACEPTIVE MOUSSE

Formula N	
Ingredients	% (w/w)
Nonoxynol 9	8.0
Lauric/Myristic Acids	2.5
Stearic/Palmitic Acids	3.5
Triethanolamine	2.2
Glyceryl Monostearate[a]	2.5
Polyoxyethylene (20) Sorbitan Mono-oleate	2.5
Polyoxyethylene (20) Sorbitan Monolaurate	3.5
Polyethylene 600 Glycol[b]	1.5
Polyvinylpyrrolidone K-30[c]	1.0
Benzethonium Chloride, USP[d]	0.2
Deionized Water	67.6
Propellant A-46	5.0

[a]Viscosity builder and foam stabilizer.

[b]Average molecular weight is about 600.

[c]Protective colloid.

[d]Benzyldimethyl [2-[2-(p.1,1,3,3-tetramethylbutylphenoxy)ethoxy]ethyl] Ammonium Chloride

TABLE 14.   MOUSSE FOR MASTITIS TREATMENT

Formula O	
Ingredients	% (w/w)
Sodium Lauryl Sulfate (30% Active Ingredient in Water) (As Duponol WA Paste, or equivalent)	27.7
Polyethyleneglycol 400 Distearate	3.5
Triton X-100[a]	0.4
Sodium Palmitate/Stearate	0.4
Sodium Sulfate 10-Hydrate	0.2
Sodium Citrate ½-Hydrate[b]	1.5
Neomycin Sulfate (as base)	3.8
Sodium Hydroxide (50% in Water) -- to pH 9.2	q.s.[c]
Deionized Water	56.5
Propellant A-46	6.0

[a]Foam modifier.

[b]Sequestrant and stabilizer.

[c]A sufficient quantity.

available in the U.S., but an interesting topical product, using dimethyl
ether as the propellant was patented about 1984. It contains chlorhexidene
acetate or gluconate plus a light-blue dye (to show the areas treated), in a
hydroalcoholic system. After milking, it is used as a germicide and the
chilling effect of evaporating diethyl ether causes the sphincter muscle to
immediately close, preventing contamination of the udder quadrants and
incidentally also preventing some leakage of milk. The product is especially
useful for older cows on automatic milking machines.

Pharmaceutical foams have been well received for over a dozen specific
applications, including hemorrhoid treatment, relief of tenesmus in deep
wounds and ear cleansing.

When foam medicinals are to be inserted into various body cavities, it is
often best to use a special type of very blunt, all-plastic syringe, such as
the model shown in Figure 1.

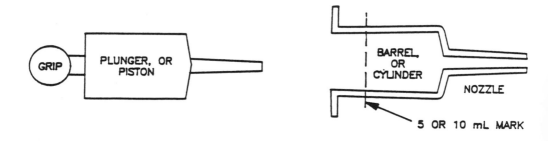

Figure 1. Foam Syringe

Directions: To use the metering syringe, fill the barrel with foam to a level
beyond the volume mark, minimizing any air pockets. Press the piston in until
it reaches the volume mark, removing excess from the nozzle tip. Insert; then
slowly press the piston in to the fullest extent. The narrow tip of the
plunger serves to minimize any wasted product remaining in the nozzle area,
since such medications are often rather costly.

Since dosage is directly affected by foam density, it is desirable to formulate a foam with a density in the range of 0.05 to 0.10 g/mL, using sufficient propellant so that the foam density at the beginning of product use will not be more than 25% less than that near the end. Uniformity of propellant fill is also very important. For example, a Pamasol "Stepped Rotary" filler contains two concentrate filler heads, a crimper, and two propellant filling heads. It will produce 35-40 meter-spray cans per minute, or 60-70 regular cans per minute, filling both product and propellant with good accuracy. It is the filler of choice (worldwide) for small, pharmaceutical products.

Shave Creams

Shaving creams--the last of the "mousse" products to be discussed--are more of a utility product than a personal care item. In the U.S., a 310-g dispenser can still be purchased on sale for less than $1.00, although most of the name-brand shave creams cost at least twice that much.

Discovered in 1931 by Eric Andreas Rotheim of Norway, who pressurized certain soap solutions with butanes, the shave cream was reborn in the U.S. in 1948. It used from 6.5 to 8.5% of a CFC-12/114 blend. In 1954, all but one marketer changed over to 3.4 to 4.0% Propellant A-46. This particular blend of propane and isobutane has air-free pressures of 3.24 ± 0.14 bar at 21°C, and 8.87 ± 0.35 bar at 54.4°C. At the time of its first production, it was the highest-pressure mixture that could be used for atmospherically crimped, standard-strength aerosol cans. Table 15 gives the formulations for three typical shave creams.

The usual shave cream formula is an 8 to 12% dispersion of sodium or potassium and triethanolamine fatty acid soaps in water, plus foam stabilizers, wetting agents, emollients, humectants, fragrance, preservatives, and sometimes special ingredients. Some fairly exotic materials have sometimes been used, such as fluoro-acrylic "super-detergents" for extra beard softening, and hyaluronic acid, a natural polymer that encourages the skin

TABLE 15.   SHAVE CREAMS

Ingredients	Formula A	Formula B	Formula C	U.S. Tradenames
Deionized Water	74.9	79.5	78.1	
Lauric/Myristic Acids	1.5	1.0	0.7	Emersol 132,et.al.
Stearic Acid (Triple X)	6.0	7.0	8.0	Emersol 655 or 621
Lauryl/Myristyl Diethan-olamine	0.5	2.0	----	Schercomid SLM-S
Sodium Lauryl Sulfate (30% Water Solution)	----	----	1.5	Duponol WAT 30%
Sodium Hydroxide (50%)	----	0.5	----	Dow Chemical Co.
Potassium Hydroxide (45%)	----	2.25	0.4	
Triethanolamine (99%)	3.9	----	3.0	Union Carbide Corp.
Cetyl Alcohol, N.F.	0.5	----	----	
Glycerin - 96%, U.S.P.	5.8	4.0	2.5	
Polyvinylpyrrolidone K30	----	0.15	----	
Mineral Oil, N.F. Grade	2.4	----	----	Witco "Carnation"
Methyl p.Hydroxybenzoate	0.1	----	0.1	
n.Propyl p.Hydroxybenzoate	0.03	----	0.04	
Fragrance	0.67	0.3	0.36	IFF #2651-AB
Lanolin Derivative	0.5	0.2	2.0	Lantrol-Malmstrom
Propellant A-46	3.2	3.1	3.3	Phillips Petroleum

Note: These formulas may be packaged in tinplate or aluminum single-lined cans.

renewal process and moisturizes the skin, to reduce the irritation of close shaving.

Formula A is the highest quality of the three, followed by Formula C. Formula A is also the most costly, but the marketer has the option of making such claims as "sodium free," "contains no alkali," or "100% organic," which may be advantageous.

In addition to the foam stability provided by the sodium lauryl sulfate (often called SLS), PVP K30, lauryl/myristyl diethanolamide (or coco-diethanolamide), and especially cetyl alcohol, formulation chemists may use Keltrol or Methocel 25MC thickeners, Monamid 150LW, Deriphat 170-C (acid zwitterion) and other ingredients. The critical test for sufficient foam stability is that the foam should not significantly drain or dry out from below the nose of the user in less than three to four minutes, even at low humidities. It should also be stiff, smooth and thick-textured, and form a 5-g puff able to support a full-length pencil at a 5° angle. Some foams are dispensed hot, either using reactive chemicals (such as sodium thiosulfate and 10% hydrogen peroxide, kept separate until the moment of use), or more commonly by using an electrical heating mantle. In such cases, the foam may be delivered "steaming hot" at up to 85°C, and quite a lot of extra foam stability agent will be required. These hot foams represent only about 4% of the 210,000,000-unit U.S. aerosol shave cream market.

The choice of propellant is critical to success. At pressures of 2.5 bar or less at 21°C, a secondary expansion will occur in the hand, which is perplexing to the user. Blends higher in pressure than Propellant A-46 will normally require a special, high-pressure-resistant can.

The amount of propellant is also important to success. The use of too little propellant results in production of a higher density foam (such as 0.12 g/mL), and the situation will worsen as the dispenser is used up. This is because emulsified propellant can escape into the expanding head space, leaving less in the liquid phase to produce the foam structure. Conversely, over 4.0% hydrocarbon propellant will produce a relatively dry, airy foam with

poor wetability and smooth-out properties. Excessive propellant may escape
during actuation, leaving the foam puff defaced, with a pock-marked, wrinkled
or roughened surface. The noise level will be higher as the propellant tears
the foam micro-structure during the escape process.

Some work has been done with alternative propellants. Those with both
oil and some water solubility produce less stable foams, such as HFC-152a, but
the stability may be adjusted by using blends of HFC-152a and hydrocarbon
propellant. Dimethyl ether will not produce a foam. Nitrous oxide generates
satiny foams with very fine micelle structures, as does carbon dioxide.
Despite the low solubility (about 1%), these foams remain reasonably uniform
throughout package life, especially if only 60-65% product is placed in the
can.

Many unusual foams have been produced. There are exploding foams,
bouncing types, crackling foams, and even anhydrous foams of astonishing
durability. Several anhydrous types are used for the application of topical
pharmaceuticals.

Several food foams have been introduced, but the only really successful
one is whipped cream, which is nearly always pressurized with approximately
1.2% nitrous oxide. The slow growth of psychrophilic bacteria, even at 2°C,
has posed a major shelf-life problem for many of these products. One U.S.
filler has developed a process for producing sterile products, but enzymes
will adversely affect even these products within a few months.

Whipped food products that have not been marketed to any extent are
pancakes (crepes), expanded mayonnaise, whipped syrups, ice cream toppings,
chocolate milk additives, and alcoholic toppings for Irish Coffee and certain
mixed drinks. In the U.S., the hydrocarbon propellants, nitrous oxide, carbon
dioxide, and nitrogen, are the only propellants permitted for food uses. In
some other countries, nitrous oxide is not allowed in food products. One of
the key problems with food aerosols is that the final dispensers cannot be
autoclaved to render the contents sterile without causing the cans to burst.

Underarm Products

In the U.S., these aerosol products fall into two distinct classes: the antiperspirant and the deodorant. Antiperspirants are considered drug products by the U.S. Food and Drug Administration because they have a definite physiological effect on the pores of the skin, and they are regulated accordingly. On the other hand, a simple deodorant consists of an alcoholic solution of a germicide, which does nothing except to reduce the population of skin resident bacteria for several hours, thus inhibiting their ability to degrade certain ingredients in natural perspiration into malodorous materials. Antiperspirants also function as deodorants, because they reduce the pH value of the skin to about 3.2 to 4.2 at 35°C, and this level of acidity inhibits the proliferation of microorganisms.

In the U.S., an aerosol antiperspirant must, by definition, act to reduce the rate of perspiration. More specifically, antiperspirants must reduce the output of the apocrine sweat glands by twenty to seventy per cent. These glands are actually located not only at the armpits, but in the ano-genital area, the eyebrows, and (in women) behind the ears. Despite the commercialization of aerosol antiperspirants in the past for more general body application, currently available products are labeled for underarm use only.

All antiperspirants have a physiological effect on the skin, but these effects vary widely with the formula, the consumer, testing conditions, time, and other factors. Sweat reductions even between the right and left underarm areas of a given person may be quite different. In 1978, the FDA issued a set of OTC (Over-The-Counter) drug regulations for antiperspirants, titled "Guidelines For Effective Testing of OTC Antiperspirants" that defined an axillary sweat reduction study (2) as a proposed Monograph. It stipulated that an aerosol antiperspirant must reduce sweat by at least 20% after both 1 hour and 24 hours from the time of application in at least 50% of a minimum 15-person target population. The test conditions were 38°C and 35% RH (Relative Humidity) during the one-hour collection periods for each day of the five-day test cycle, and 22.2°C otherwise. Application time was described as a two-second spray under each arm.

Until the introduction of the more powerful aluminum chlorohydrate complexes in 1986, most aerosol antiperspirants fell into the 22-33% sweat reduction category. By comparison, stick antiperspirants exceed 40% and some roll-ons reach 65%. However, a sweat reduction of 25% or more represents such a dramatic improvement in the control of wetness that many users are quite satisfied with aerosol performance. During the recent introduction of the new, more potent antiperspirant materials, however, the suppliers stated that they made these changes because of some user dissatisfaction with product effectiveness, resulting in a sales reduction of the aerosol category over the years. Aerosol antiperspirants have also had modest problems with "bounce off," resulting in some nasal irritation, higher packaging costs, and occasional valve plugging. Recent attempts to reduce ground-level ozone (smog-related) have caused at least one state to study aerosol VOC emissions.

During 1985, the Reheis Chemical Company (U.S.) announced their new line of REACH antiperspirants for aerosols and other dispensing forms. These are up to 50% more potent than the standard aluminum chlorohydrate complex; e.g., $[Al_2(OH)_5]_{13} \cdot 2.5H_2O$--which has an Al:Cl ratio between 2.1 and 1.9 to 1.0, and an FDA permit level of up to 25% of the total product.

The new REACH 101, 201, and 501 compositions represent commercial forms of aluminum chlorohydrate polymer that can be separated out of the mixture of polymers in the parent compound using liquid chromatography. The Al:Cl ratio is maintained. With the greater effectiveness (up 30 to 50%, depending on choice) formulators had the following options:

- Use the new compounds at the former levels, for a 30 to 50% more effective product; or

- Use reduced concentrations of the new compound, lowering the effectiveness of the product to the former level, but reducing formula cost and bounce off.

Table 16 compares some representative antiperspirant product efficiencies.

In one form or another, aluminum chlorohydrate is the only active material now used in aerosol products, but the specific polymeric compositions and their particle size distributions have gone through a transition from 1986 through 1989 that now suggest that the REACH compounds have captured at least 80% of the U.S. market volume.   In 1988, the aerosol underarm products sold increased from 114 million units to 144 million, or 26%, although the aerosol industry grew by a comparatively small 6.8 percent.   This extraordinary growth is said to have been the result of the availability of more effective products and well-advertised introductions by Mennen, Bristol-Myers, and other marketers.   The aerosol share of the underarm market also grew, from 31% to 37%.

The present products contain from 7.0 to 12.5% of various aluminum chlorohydrate complexes.   Without the addition of a dispersing system, the suspended aluminum chlorohydrate would settle into a bottom layer between uses and would require long and difficult shaking to redisperse.   Also, the material within the very bottom of the dip tube will not redisperse even with shaking, leading to valve plugging and consumer dissatisfaction.   Because of this, all aerosol antiperspirant formulas contain a dispersing agent, normally a surface-treated form of montmorrillonite clay activated by the inclusion of a relatively polar ingredient such as ethyl carbonate or ethanol.   With the correct combinations and balances and good compounding techniques, clogging problems can be avoided.

Table 17 presents the formulation of a low-cost, low "bounce off" antiperspirant of above average effectiveness.

Both the isopropyl myristate and silicone fluids of the formula shown in Table 17 are useful as inert carriers to transport the aluminum chlorohydrate to the skin surface and stick it there with a minimum of bounce off.   These ingredients are also needed as slurrying material for the aluminum chlorohydrate to facilitate the production process.

TABLE 16.   COMPARISON OF ANTIPERSPIRANT EFFICIENCIES

Generalized Composition and Form	% Efficiency	
	First Day	Fourth Day
3.5% Old CFC Formula (Banned in 1978)	25	
10.0% Hydrocarbon-Type Aerosol	25 ± 7	29 ± 7
12.5% Hydrocarbon-Type Aerosol	30	
10.0% Hydrocarbon-Type Aerosol (REACH 101)	44 ± 8	57 ± 6
10.0% Hydrocarbon-Type Aerosol (REACH 201)	38 ± 8	50 ± 9
10.0% Hydrocarbon-Type Aerosol (REACH 501)	35 ± 7	45 ± 10
20.0% Suspension Stick - Standard	40	
20.0% Suspension Stick - Rezal 36GP Active	55	
25.0% Roll-On Suspension - REACH AZP-703	62	

TABLE 17. AEROSOL ANTIPERSPIRANT

Ingredients	% (w/w)
Aluminum Chlorohydrate (REACH 101)[a]	8.00
Quaternium 18 Hectorite (Bentonite 38)[b]	0.82
S.D. Alcohol 40-2 (Anhydrous)	0.80
Dimethylsilicone [500 centistokes (cstks)][c]	1.50
Isopropyl Myristate[d]	1.00
Cyclomethicone F-251[e]	7.63
Perfume Oil[f]	0.25
Propellant A-31 or A-46[g]	80.00

[a]REACH 101 is produced by the Reheis Chemical Company, 235 Snyder Avenue,
Berkeley Heights, NJ. (201)464- 1500. (An equivalent product is produced by
the Wickhen Division of the Dow-Corning Corporation.)

[b]Bentonite 38 is a surface quaternized form of montmorillonite clay, offered
by NL Industries, Inc. as "Bentone 38."

[c]Dimethylsilicone (500 cstks.) is available from Dow-Corning Corp., General
Electric Silicones Division, and others. Purchase the anhydrous (clear)
liquid, not the emulsion forms.

[d]Isopropyl Myristate is sold by Van Dyk & Co., Inc. and many other firms.

[e]Cyclomethicone F-251 is a physical mixture of Cyclomethicone D-4 Tetramer
25%, and Cyclomethicone D-5 Pentamer 75%. It is available from the Dow-
Corning Corp.

[f]The perfume oil is the marketer's choice. It must be purchased from a
perfume house that is advised that it will be used in a REACH 101 aerosol
antiperspirant, and then tested for compatibility with the product. The
percentage may vary, according to fragrance intensity and the marketer's
preference.

[g]Propellant A-31 (with vacuum crimp) is preferred, giving a can pressure of
about 2.5 bar at 21.1°C. The use of Propellant A-46 will provide an initial
pressure of about 3.6 bar at 21.1°C (with vacuum crimp) and thus a delivery
rate approximately 19% faster at the beginning of use. Preferential removal
of air and propane through the vapor-tap orifice will reduce pressure to
about 2.6 bar at 21.1°C when the dispenser is 50% empty, and to about 2.3
bar at 21.1°C at the point of incipient emptiness.

(Continued)

TABLE 17.   (Continued)

Notes:

1.   Concentrate Viscosity: 18,000 cps. (Brookfield Viscometer; 10 rpm.)

2.   Concentrate pH Value: 3.3 - 4.7 (1:10 v. dilution in deionized water)

3.   Concentrate must be homogenized to remove probable clumps or clusters of oil-wetted aluminum chlorohydrate powder.

4.   REACH 101 is hygroscopic and must be protected from the moisture in ambient air at all times. Once it is added to the oily concentrate it will no longer absorb moisture. The following recommendations apply:

   •   Keep drums closed except when sampling or using;

   •   Do not remove packets of moisture-absorbent silica gel in drums, if present, except when adding REACH 101 to batch;

   •   Add REACH 101 as quickly as practical, while still minimizing clumping; and

   •   Rinse off tank walls of REACH 101 powder, immediately after addition.

5.   If a U-t-C (Under-the-Cap) gasser is used, the last 10% of the propellant must be added by means of a following T-t-V (Through-the-Valve) gasser in order to clear thick concentrate from the dip tube and avoid possible valve-plugging problems.

6.   All equipment touching the concentrate should be of 304 or 316 stainless steel, Tygon® hose, or approved rubber hoses in good condition.

7.   The concentrate is sufficiently thick or viscous that settling of the solids will be a slow process. However, continuous recycling of the concentrate is required, as well as some slow stirring of the concentrate filler bowl. Over-circulation, such as short-loop circulation during filler downtimes, is not recommended. It may cause localized heating and micro-splintering of the solid aluminum chlorohydrate.

The isopropyl myristate improves feel (or handle) and texture. Being essentially nonvolatile, it helps the Dimethylsilicone (500 cstks.) carrier keep the underarm area lubricated for many hours.

The other silicone materials are relatively volatile. The F-251 blend is selected for a combination of effectiveness and lower expense. They are used to impart an extra measure of carrying ability and skin lubricity, preventing any tackiness development as the aluminum chlorohydrate slowly dissolves in perspiration films. Having done their work, they slowly evaporate, preventing long-term excessive oiliness and the staining of clothing in the underarm area.

The isopropyl myristate, and various silicone fluids also help the operation of the aerosol powder valve. They reduce the amount of bulking agent needed to prevent hard-packing of solids between product applications. Finally, they provide spreading characteristics that help distribute the aluminum chlorohydrate more effectively across the dermal surface.

If excessive amounts are used, the oil may coat the aluminum chlorohydrate so effectively that it cannot contact skin moisture and begin dissolving. This will cause a lag between the time of application and the time antiperspirancy becomes apparent. Fabric staining, "wetness," and cost will all increase if too much isopropyl alcohol and silicone fluids are used.

Other materials have been suggested as replacements for these ingredients. For example, Croda, Inc. suggests replacing isopropyl myristate with their Procetyl AWS (a propoxylated/ethoxylated ether of cetyl alcohol). Union Carbide suggests using Fluid AP, and others have promoted such items as myristyl myristate ester and octyl palmitate ester. Such ingredients are added at some risk. For example, some samples of myristyl myristate contain small amounts of unreacted myristic acid, which can seriously reduce or even eliminate the antiperspirancy of the aluminum chlorohydrate.

The Bentone 38 surface-polarized montmorillonite clay is very finely divided and has the approximate formula $NaCa[(Al,Mg)_2Si_4O_{10}] \cdot Q \cdot nH_2O$, where Q is a specific quaternary ammonium compound designed to increase surface charge and further promote suspending properties. A small amount of ethanol is added to activate the polar surface and augment charge separation.

When the balance is achieved, the Bentone 38 and ethanol system will slow down the settling of the aluminum chlorohydrate and allow it to eventually settle into a lose, voluminous layer between product uses. A gentle inversion or shake will then swirl the solids back into suspension. This should be checked for any new formula (even a new perfume in a tested formula) using glass compatibility equipment. If the aluminum salts can settle into a hard, obdurate mass--regardless of settling rate--the product will suffer from problems of reconstitution and probable valve plugging.

Both scented and unscented antiperspirants are marketed. The unscented versions are sometimes preferred by men, and always by hypoallergenic persons. They are usually very lightly scented, despite the label, using approximately 0.04% of nondescript perfume oils to cover the slight chemical odors of the other ingredients. Some products contain encapsulated perfumes, in addition to the "non-encap" perfume. They provide longer-term fragrance release, as moisture dissolves the modified polyvinyl alcohol (0.001"(0.025mm)-diameter) micro-capsules of additional fragrance.

The aluminum chlorohydrate easily develops enough acidity under the arm to prevent the proliferation of skin-resident, odor-causing bacteria. Thus, no special microbicides need be added, except to treat cases of chronic hyperhydrosis or certain other dermal pseudomorphoses.

The homogenization step for the concentrate-processing stage should be a one-stage, rather gentle one designed to break up clusters of oil-saturated aluminum chlorohydrate, rather than to fracture the roundels of the salt itself.

The roundels are already sufficiently fine-particled that they will not clog an aerosol valve. But if they are broken up into a lot of "splinters," buildup and possible clogging might occur.

Sample amounts of the concentrate should be made in the laboratory so that the viscosity and flow characteristics of the batch-making process can be considered. A variable speed, planetary, top-entering agitation system is needed for best results. When the powder is added and viscosity increases, all parts of the mixture must be agitated.

The compounding procedure is as follows:

- Add the isopropyl myristate, through a 5-micron filter;

- Add the Cyclomethicone F-251, through a 60- to 100-mesh screen;

- Begin agitation at about 75 rpm;

- Begin recycling, out the bottom and back into the tank via a pipe that extends to the lower one-third, to prevent splashing. The pump in this system should be set at around 150 to 200 rpm to prevent shearing. A Cuno or similar filter in this line will be bypassed during the compounding stages;

- Pre-weigh the Bentone 38 and add manually to the tank at a fairly slow rate, about one 20-kg bag per minute at most;

- Agitate at least 15 minutes, until the Bentone 38 is dispersed;

- Add the SD Alcohol 40-2 (Anhydrous);

- Operators in face masks and protective clothing then add powdered aluminum chlorohydrate REACH 101 to the mix-tank at a rate of about 25 to 50 kgs per minute:

    -- Build batch size around full-drum amounts of REACH 101 if possible, to prevent dealing with partial drums,

--   Protective masks and clothing are needed because of the genera-
        tion of irritating dusts during additions;

- Rather quickly, add the dimethylsilicone (500 cstks) to the batch
  tank, using a suitable Tygon or rubber hose, so that deposited
  powder on the walls and dome of the mix-tank can be rinsed down into
  the batch;

- Add Perfume Oil;

- Agitate at least 15 minutes. Check for dispersion uniformity;

- Arrange recycling line to pass product through stainless steel Cuno
  filter (coarse) and through either a Votator or Homogenizer to
  homogenize the lumps. Use a filter with about 0.13-mm spacings.
  Use a flow rate of 150 to 200 kg per minute;

- Pass the finished concentrate into a stainless steel, agitated
  holding tank, with a recycling line as close to the concentrate
  filler as practical. The temperature will have increased some 5° to
  10°C during rotating or homogenizing, but it should not be allowed
  to be over 40°C.

The valve is often supplied by Precision, Seaquist, Valois, Aeroval,
Summit, or other major manufacturers in a rounded edge, powder-valve design.
In the U.S., a large, 20-mm diameter, white, one-piece button is used. It is
ordered separately from the valve and applied by hand (rarely) or by an
automatic button tipper, often to line up with the 180° reverse-directional
dot on the crown of the valve cup.

A prototype valve is one with a 0.46-mm stem orifice, 0.63 x 0.46-mm
vapor-tap body, neoprene gasket, and 0.50-mm straight-bore actuator button.
The delivery rate can be changed downward if desired by using a 0.50-mm vapor-
tap orifice instead of the 0.46-mm size. Many formulators check several
larger-volume products on the market and decide which ones have the best spray

pattern, delivery rate, and other characteristics they require. They then
contact the appropriate valve company, identifying the product, and ask for a
sample valve made to the same specifications. Unless it is a customized
component (which is rare, except perhaps for color) the valve-maker will
always comply.

Two forms of the REACH 101 antiperspirant powder, differing only in
particle size distribution, have been used in aerosols:

- MICRO-DRY "REACH 101" (Reheis): Standard. Impalpable. More than
  99.8% of the particles are smaller than 74μ;

- MICRO-DRY "REACH 101" (Reheis): Ultrafine. At least 99.8% of the
  particles are smaller than 50μ.

A filtration step, using a Cuno or equivalent cartridge filter of 0.13-mm
retention, removes agglomerates, oversized particles, tramp cellulose fibers
(from bags), and other extraneous solids from the finished batch of concentrate. Experience suggests that the Standard, Impalpable grade is quite
satisfactory and somewhat less of a potential problem in terms of the rate of
moisture pickup during handling. However, it is always a good idea to contact
the suppliers (Reheis and Dow-Corning Corporation, in the U.S.) and ask for
recommendations and literature.

The aerosol can may be a necked-in 200-201/202x406 or 509 or 514-mm can
in the U.S. market, which is equivalent to a 51-51/52x111 or 140 or 148-mm can
elsewhere. The necked-in version is a marketer preference, based mainly on
aesthetics, not on technical or functional requirements. The so-called
"straight-wall" cans are also acceptable. Because the low-density hydrocarbon
propellants are nearly always used at levels approximating 75%, it is
customary to place a 100-to 115-gram fill weight into the can size just
described.

In the U.S., the necked-in cans have a 50.70 $\pm$ 0.25-mm industry specification for the diameter of the top double seam. This relatively difficult

specification reflects a need for uniformity to prevent full-diameter straight-wall cover-caps from fitting too tightly or too loosely when applied over the seam.

The cans are generally of minimal tinplate throughout, with a single lining and usually a stripe over the welded side seam. Adherence to Good Manufacturing Practices (GMP), which includes code legibility, is a general requirement for this Over-The-Counter drug product.

The personal deodorant (or underarm spray deodorant) complements the antiperspirant. It is often used by those who have either constant or sporadic skin irritation problems with antiperspirants because of their salinity, astringency or acidity, or by those for whom underarm perspiration is either not a problem, or cannot be controlled by antiperspirants because of the environment or type of activity.

The basic personal deodorant formula consists of a germicide, fragrance, ethanol solvent, deionized water (sometimes) and propellant. The standard propellant in the U.S. is either isobutane, or a mixture of up to 37 wt% propane in isobutane. In Europe, both hydrocarbon and dimethyl ether propellants have been used. Various HFC and HCFC propellants could technically be used, but their higher cost has so far effectively precluded their use. Four representative formulations are given in Table 18.

The size of the personal deodorant market is now about 55,000,000 units in the U.S., 34,700,000 in 1988 in Japan, and relatively small in Europe. The containers are similar in size and logo to the antiperspirant aerosols, and are sometimes purchased by mistake because of this. Most major marketers offer both products in two or three sizes and in both scented and unscented

TABLE 18.   PERSONAL DEODORANTS

Ingredients	Formula A (%)	Formula B (%)	Formula C (%)	Formula D (%)
Irgasan DP-3000[a] Germicide	0.11	----	----	0.12
Genzthionium Chloride[b]	----	0.20	----	----
Methyl p.Hydroxybenzoate	----	----	0.03	----
n.Propyl p.Hydroxybenzoate	----	----	0.06	----
Benzyl p.Hydroxybenzoate	----	----	0.08	----
Propylene Glycol, U.S.P.	1.50	----	1.03	----
Dipropylene Glycol	----	1.05	----	2.00
Zinc Phenolsulfonate[c]	----	----	----	1.00
Fragrance	0.35	0.25	0.30	0.38
S.D. Alcohol 40-2 (Anhydrous)	58.00	68.50	63.50	13.35
Deionized Water	----	----	----	47.00
Sodium Benzoate	----	----	----	0.15
Isobutane (A-31)	40.00	----	----	----
Propellant Blend A-46 16 wt% Propane in Isobutane	----	----	35.00	----
Propellant Blend A-70 37 wt% Propane in Isobutane	----	30.00	----	----
Dimethyl Ether	----	----	----	36.00

[a]2,4,4'-Trichloro-2'-hydroxydiphenylether.

[b]Benzyldimethyl [2-[2-(p.1,1,3,3-tetramethylbutylphenoxy)ethoxy]ethyl] Ammonium Chloride.

[c]Zinc Sulfocarbolate.

versions. The packaging requirements for the hydrocarbon-propelled formulas
are designed to give a fairly fine-particled, low delivery rate spray, using a
vapor-tap valve with (typically) a 0.33-mm diameter orifice and a mechanical
break-up button. For the dimethyl ether products (as in Formula D of Table
18), very efficient valves are required to break up the large amount of water
present. The Precision Valve Corporation's 2 x 0.50-mm "Aquasol" stem valve,
0.50-mm MBST (Mechanical Break-Up, Straight Taper) button, and butyl rubber
stem gasket valve may be used. The supplier should be contacted for specific
recommendations, but sprays of 60μ average particle size are obtainable. The
somewhat higher cost of dimethyl ether in most areas can be justified by its
ability to incorporate significant amounts of water in solution, giving the
feeling of excessive wetness. The flame projection of this formula will vary
somewhat with valve selection, but it is generally a 100- to 150-mm small,
sputtering light blue plume or flare.

Colognes and Perfumes

A cologne is generally considered to be a dilute form of the perfume or
sachet product, containing from 1.5 to 6.5% of essential oil or fragrance
compound. The true perfume may contain from 6.5 to 14.0 percent. The carrier
is almost always ethanol, generally anhydrous ethanol, and the propellent is
often a hydrocarbon type. For perfumes, which are smaller and more costly
than colognes, the glass or aluminum dispenser carries a meter-spray valve
able to dispense about 0.05 gram per actuation. A typical 20-gram fill will
offer about 400 actuations to the user.

The European innovation known as the deo-spray is a form of cologne and
has the typical cologne composition. No deodorant is present, as might be
inferred from the generic name. Packaged in lined aluminum cans holding as
much as 200 grams, the deo-spray or deo-cologne provides an inexpensive option
for spraying one's skin or clothing with a relatively low-cost but still
acceptable fragrance. Some deo-sprays are made especially for use by younger
children and are labeled accordingly. Plastic caps resembling flowers and
animal heads have been used for added appeal. Finally, the sachet spray falls
somewhere between a cologne and a perfume. It contains approximately 4 to 8%

fragrance compound formulated to as light a color as possible to prevent the staining of lingerie, handkerchiefs, and other fabrics. Two typical cologne formulas are presented in Table 19.

Glass has been the accepted standard for colognes since the origin of the aerosol cologne in 1953. While plain glass containers of up to 125-mL capacity have been marketed, glass bottles of greater than 30-mL capacity are usually plastic coated. In the U.S., the Wheaton Aerosol Company's "Lamisol" container is often used. It has a vinyl-based covering that firmly adheres to the glass surface, making it more resistant to fracturing if dropped and helping to contain glass fragments and flammable vapors if the container does break. At present, the U.S. Department of Transportation limits the size of non-metallic aerosol dispensers to a capacity of 118.3 mL (without a special exemption).

The use of the OPET, Petasol, and similar plastic bottles is being studied in the U.S., Japan, and Europe. Based on biaxially-oriented polyethylene terephthalate, these bottles offer lightness, great break resistance, clarity, translucency, or opacity, and more freedom of shape and design than glass. Their very lightness has been viewed as a marketing deterrant, since buyers are accustomed to the solidity and weightiness of glass as a quality attribute. These bottles cannot be used for strong solvents, such as dimethyl ether, since they lose strength rather rapidly at temperatures of over 60°C, and may suffer from permeation effects when used with some formulations. In England, hair sprays (not too different from many cologne formulas) have been marketed successfully as 200-gram fills in OPET bottles.

The levels of HFC-152a ($CH_3 \cdot CHF_2$) or HCFC-22 ($CHClF_2$) used in Formula A in Table 19 are considered minimum levels. The vapor pressures of both propellants are suppressed when they are added to ethanol, an effect reduced when water is incorporated as a third ingredient. This is why Formula A contains 13% deionized water. An aerosol valve with maximum breakup power is used for all cologne formulations.

TABLE 19.   COLOGNE FORMULATIONS

Ingredients	Formula A (%)	Formula B (%)
Fragrance	4.00	4.00
Di-n.butyl Phthalate	2.00	- - - -
Sodium Saccharinate[a]	0.01	- - - -
FD&C and/or D&C Dye Solution[b]	0.09	- - - -
S.D. Alcohol 40 or 39C (Anhydrous)	65.00	76.00
Deionized Water	13.00	- - - -
HFC-152a or HCFC-22	15.90	- - - -
Isobutane A-31	- - - -	20.00
Packaging mode:	Glass	Aluminum

[a]The Sodium Saccarinate (or a similar synthetic sweetener) is added to nullify the rather tart bouquet of the ethanol.

[b]In the U.S., these are Food, Drug, and Cosmetic-approved dyes or Drug and Cosmetic-approved dyes in the form of a stock solution of various concentrations.   The solvent is generally deionized water, but propylene glycol and/or a preservative may be added as well.

Perfumes, deo-sprays, sachets, and related products have relatively similar formulations. Fragrance is deposited on an animate or inanimate surface with high efficiency. The spray must be relatively wet, but still retain cosmetic elegance.

Although dimethyl ether could be used in fragrance products, and there are those who claim it imparts a cleaner, fresher odor, especially when water is included, others suggest that the propellant itself has a stronger odor than the purified hydrocarbons, and certainly much higher than HFC-152a, which has almost no odor at all. Fillers in various parts of the world continue to use CFC-12/114 blends for perfumes and colognes, partly because of the greater importance of reduced final-product flammability when dealing with a frangible material such as glass. Some of these fillers are not set up yet to safely handle flammable propellants. In time, they will have the option of converting to nonflammable HCFC-22 propellant or to one of the nonflammable future alternative types, such as HFC-134a. The latter is a modest solvent and is virtually odorless.

Perfumes are extremely complex mixtures of both natural and synthetic materials, and it is rare that all of them are soluble in the complete aerosol formula. In some instances, these resinous substantives will only precipitate after several days or weeks. When they do, they usually agglomerate into fairly hard masses, readily capable of causing sputtering, distorted spray patterns or even plugging the aerosol valve. In a clear glass container the precipitation can be seen, and it gives a very negative image of product quality.

An early industry practice was to store the complete formulation in a covered 2,000-liter tank at -10°C for several days, then to filter out the dregs enroute to the product filler. With the flammable propellants, this technique is no longer practical, although the complete concentrate can certainly be held for a time at room temperature and then filtered free of precipitates. The perfume suppliers are well aware of the problem and can sometimes provide fragrances that have been tested to show that the addition of hydrocarbon propellants to the filtered concentrate will not cause any

further precipitation. The marketer should always make the "two-week test,"
which is to prepare the finished formula in either a clear glass aerosol, or
in a clear glass product-compatibility tube of about 100-mL capacity, holding
it for one week at 35 to 40°C, and then for one week at 2 to 4°C. The sample
unit is then evaluated for clarity or haziness, for precipitation, and for
good odorous stability when compared with a freshly prepared standard in
glass. The darker-colored fragrance products seem to be more prone to
precipitation.

When fragrance products are packaged in small aluminum cans, the cans
should be lined with an epon-phenolic or similar material. Otherwise, the
bare aluminum metal may have a reducing effect on aldehydes and certain other
sensitive perfume ingredients. Some perfume components are known to cause or
enhance corrosion reactions, especially the citrus types, such as bergamots
and citronellal bases. Test packing is essential. Formulations that contain
water are especially critical.

HOUSEHOLD PRODUCTS

General Comments

The "household products" segment of the U.S. aerosol industry reached a
total of 1,424,100,000 units in 1988, accounting for 48.8% of the aerosol
industry total and making this the largest category. See Table 20 for
details. In Japan, the same segment amounted to 123,554,000 units in 1988,
amounting to only 25.5% of the industry total. In that country, the largest
market share of aerosols (44.3%) is held by personal care products.

In the U.S., the term "household products" includes all consumer aerosols
except for pesticides, personal care items, and foods and drugs. They are
administered by the Consumer Product Safety Commission (CPSC), which is a
federal agency created in 1972 to handle the Federal Hazardous Substances Act
of 1960, the Poison Prevention Packaging Act of 1970, and other laws. Among
other things, they recommend pre-market testing of aerosols for flammability,
inhalation toxicology, skin and eye toxicology, ingestion toxicology, and

TABLE 20.   HOUSEHOLD AEROSOL PRODUCTS SOLD IN THE U.S. DURING 1988

Product Type	Number of Units	% of Total
Paints, primers, and varnishes	306,300,000	10.5
Paint strippers, "snow," decoratives	24,500,000	0.8
Room deodorants and disinfectants	181,200,000	6.2
Cleaners (glass, oven, rug, tile)	167,200,000	5.8
Laundry products (starch, pre-wash)	146,000,000	5.0
Waxes and polishes	129,100,000	4.4
Other (shoe polishes, anti-static spray)	26,800,000	0.9
Refrigerant & air/conditioner refills	90,700,000	3.1
Windshield & lock de-icers	5,700,000	0.2
Cleaners (automotive upholstery)	16,300,000	0.6
Engine degreasers	27,000,000	0.9
Lubricants and silicones[a]	92,900,000	3.2
Spray undercoatings	15,700,000	0.5
Tier inflator & sealants	34,400,000	1.2
Carburetor & choke cleaners	57,100,000	2.0
Brake cleaners	29,700,000	1.0
Engine starting fluid	30,900,000	1.1
Other automotives & industrials	42,600,000	1.5
	1,424,100,000	48.9[b]

[a]Penetrating oils, demoisturizers, rust-proofers, mold release agents, tablet machine lubricants, etc.
[b]The U.S. 1987 figures were 1,326,000,000 and 48.7%.

Note: During 1988, household aerosol products increased by 7.4% in unit volume, compared with a 6.8% growth of the aerosol industry as a whole.

(sometimes) dermal corrosivity. Additional clinical studies may be needed in
some cases. If the studies are not run, or if the overall results are not
placed on the label in a prescribed format, the agency will impose severe
sanctions on marketers whose products are found to be injurious to consumers.
In addition, any torts (lawsuits) will be much more readily prosecuted by
plaintiffs against marketers whose product labels are found to not meet
federal standards.

As with all U.S. aerosol products, the primary content declaration must
be in units of weight (Avoirdupois ounces and pounds), although a volume or
weight subsidiary declaration in the metric system is acceptable. The size
of various signal words, statements of hazard, precautions, directions, weight
declaration and other informational statements is controlled by CPSC regulations.

Most household aerosol products consist of a dispersion of solids or
liquids in a continuous liquid phase. For paints, a group of finely divided
pigments is suspended in a resin/solvent/propellant solution. For starches
and fabric finishes, as well as cleaners, a colloidal suspension or emulsion
of various organic materials is prepared in an aqueous solution. These
products are dispensed in various ways, as shown in Table 21.

Water-out emulsions are used for most cleaners, but oil-out types are
used for air fresheners, the foam-type charcoal lighters with approximately 5%
water, and certain other products. The choice of propellants is very broad.
In addition to all the propellants described above, isopentane (boiling point
= 29.8°C), helium, oxygen, and even 0.2μ filtered compressed air have been
used for a few specialty items. In the U.S., CFC propellants may be legally
used for the product types shown in Table 4. Sweden, Norway, and Austria, for
example, have much shorter lists of exempted or excluded products, while most
other countries currently have either a production/importation restriction, or
no limitations at all.

In contrast with personal care products, pesticides, and most others,
household products often have very low quantities of propellant in the

TABLE 21.   HOUSEHOLD AEROSOL PRODUCT DELIVERY MODES

Product	Delivery Mode
Air Fresheners	Spray
Hard Surface Cleaners	Foaming Spray
Foam-Type Charcoal Lighters	Foam
Lubricants; Decorative Strings	Stream
Boat Horns; Electronic Cleaners	Gas
Silica-based Absorbent Powders	Liquid/Solid Spray
Caulking Compounds	Paste
Lithium Stearate Grease	Gel
Talc-based Lubricants; Wind Direction Indicators for Golfers	Powder

formulas. For example, window cleaners often have 3 to 4%, starches may have 5 to 6%, certain heavy-duty cleaners may have 6 to 8%, and rug or upholstery shampoos usually carry 8 to 10 percent. The minimum amount is determined by the following factors:

- Gassing machine accuracy;

- Propellant seepage out of the dispenser during its shelf life;

- Propellant separation from the product during use, going into the expanding head space to try and maintain pressure;

- Propellant discharge during use, because of its slight solubility or entrainment in the concentrate; and

- Consumer misuse, causing momentary release of propellant phase through the valve.

Perhaps the lowest level of hydrocarbon propellant was 1.8% n.butane, used for a low-foaming window cleaner. After several years, the pressure and use level were increased. As a general rule, the amount of nitrogen or compressed air that can be pressure-filled into an aerosol can is about 1 gram per 100 mL of capacity. At levels much above this, the pressure becomes excessive.

Household aerosol products have a greater history of consumer complaints than do other aerosols. This is because they have longer shelf and service lives, often contain more powerful solvents, are stored in a greater diversity of places and conditions, and are sometimes deliberately misused. Examples of misuse are painting graffiti and the deliberate concentration and inhalation of paint vapors and other aerosols as well. In the U.S., the "Extremely Flammable" label seems to be limited to household aerosols.

The conditions of use have a profound effect on the degree of flammable hazard to the consumer or his property. Paints should only be used where adequate ventilation is available. A concrete block moisture sealer was banned by CPSC in 1974 because it was 100% flammable in composition and two to three large cans were used at a time for basement waterproofing purposes. Several lives were lost, and over a dozen houses burned down.

A fabric protectant product, designed to spray a fluoroacrylic oil and water resistant film onto entire upholstered sofas and chairs, could not have been responsibly marketed in flammable form (using acetone or ethanol as the solvent, for example). This product has the following formula:

UPHOLSTERED FURNITURE STAIN-GUARD SPRAY

3%	Fluoroacrylic or other stain-repellant active ingredient
1%	n.Butyl Acetate (Extender)
91%	1,1,1-Trichloroethane - Inhibited
5%	Carbon Dioxide

Since 96% of this formula is nonflammable, it has enjoyed great success in an important niche area; however, it has an Ozone Depletion Potential (ODP) of about 0.15. (The ODP is a relative index of ozone destruction efficiency; the value reflects the atmospheric lifetime and the chlorine content of the molecule.)

Household products have pressures that vary from about 2 to 8 bar at 21.1°C, which is the equivalent of 6.7 to 12.7 bar at 54.4°C. Those formulas that have the higher pressures often use nitrogen, nitrous oxide, or carbon dioxide as the propellant and are designed for use at very low temperatures (such as -10°C). Applications include windshield de-icing, engine starting, and dispensing a non-slip surface on ice for cars stuck in snow or ice. These gases will still retain about half their room-temperature pressure when the dispenser is chilled to -10°C, whereas most of the other propellants will sink down to such low pressures that the products become essentially unusable.

Window Cleaners

The window cleaner, developed in 1954, was the first of a large array of water-based cleaning sprays, such as hard surface cleaners, whitewall tire cleaners, oven cleaners, bathroom (basin, tub, and tile) cleaners, and laundry cleaners for spot application to difficult stains on textiles before general cleaning in the washtub or washing machine.

Window cleaner products are water solutions with from 5 to 12% hydroxylic solvents, to which a very small amount of detergent is added. Iso-propanol (C_3H_7OH) is nearly always used because it is a good grease solvent and its odor is associated with cleanliness. Additional odorants are optional. Some marketers prefer to add ammonium hydroxide (NH_4OH) in concentrations of up to 0.3% of the commercial 29% solutions of ammonia (NH_3) in water. The ammonia actually does little cleaning, but consumers associate its odor with cleanliness. Sometimes a minor amount of fragrance may be included. Oily materials and excess amounts of detergent must be avoided, or a film may be left on the glass surface, giving a halo effect in some situations. The percentage of propellant in the formula is usually 3.2 to 5.0% isobutane.

The organic grease-cutting solvents may include butoxyethanol (C_4H_9O-CH_2OH), isopropanol (C_3H_7OH), propylene glycol monomethyl ether [$HOCH_2$-$C(CH_3)H$-OCH_3), and propylene glycol monobutyl ether [$HOCH_2$-$C(CH_3)H$-OC_4H_9]. The ratio is about two parts of isopropanol (C_3H_7OH) to one part of one or two of the solvents. The detergent selection is more critical to success.

Apart from the detergent benefit, a certain amount of foam structure is needed to show where the product has been applied and also to prevent dripping from vertical surfaces. If the foam is too voluminous or stable, the wiping cloth will simply push it around, without removing it by absorption. Also, too little detergent will reduce cleaning action, while too much will cause streaking on the cleaned glass surface. Some typical window cleaner formulations are shown in Table 22 and two specialized glass cleaner formulations are shown in Table 23.

TABLE 22.   WINDOW CLEANER FORMULATIONS

Ingredients	Formula A (%)	Formula B (%)	Formula C (%)
Isopropanol - 99%	4.0	5.0	4.0
Propylene Glycol Monoethyl Ether	3.0	2.5	----
Butoxyethanol	----	----	2.0
Sodium Lauryl Sulfate[a]	----	0.2	----
Lauryl Di-isopropanolamide	----	0.1	0.1
Ammonium Lauryl/Myristyl Alcohol EO 3:1 Sulfate	0.2	----	0.1
Sodium Nitrite	0.1	0.2	0.1
Ammonia (29% NH_3 in Water)	0.2	0.2	0.2
Deionized Water	89.0	88.5	90.0
Isobutane A-31	3.5	3.3	3.5

[a]Such as Sipon WD, a product of the American Alcolac Corp.

TABLE 23.   ANTI-FOGGING OR ANTI-STATIC GLASS CLEANER FORMULATIONS

Ingredients	Formula A (%)	Formula B (%)
Dioctylester of Sodium Sulfosuccinic Acid[a]	0.05	0.08
Silicone Glycol Copolymer[b]	0.30	0.40
Alkoxylated (8) n.nonylphenol[c]	0.10	----
Propylene Glycol Monoethyl Ether	3.52	3.49
Isopropanol - 99%	8.00	10.00
Morpholine	0.03	0.03
Deionized Water	84.00	82.00
Isobutane A-31	4.00	4.00

[a]As Aerosol OT-100, by the Chemical Product Division of the American Cyanamid Company, or Monawet MO-70E by Mona Industries, Incorporated.

[b]As Dow Corning 193 Surfactant, by the Dow-Corning Corporation. Water-soluble, gives gloss, non-tackiness, anti-fog, surface tension depression, and anti-static properties.

[c]As Triton W-30 by the Rohm & Haas Company, gives added grease removal and cleaning power.

The products in Table 22 are used mainly for windows, but they can also be employed to clean refrigerators, stove tops, kitchen counter tops and other hard-enameled, painted, or chinaware surfaces. Because of their special properties and somewhat higher cost, the aerosols in Table 23 are used more for bathroom mirrors (to prevent steaming), where anti-static properties are desired, and where a relatively glossy, polished appearance of the glass surface is desired. One version of this type of product is eyeglass lens cleaner, with dispensers in the 15-gram size that use metered spray valves delivering approximately 50 microliters per actuation. These products give about 300 actuations. Despite a general preference for the small sizes, containers holding as much as 460 grams of "Lens Cleaner - Antifog" are on the market.

Spray Starch

The self-pressurized starch was developed around 1958 and used about 4% of highly refined corn starch as the essential ingredient. Some of these starches can be dispersed into the aqueous phase at temperatures as low as 30 to 35°C, but others may require pre-cooking a 20% starch/borax (Sodium Tetraborate 10-Hydrate; $Na_2B_4O_7 \cdot 10H_2O$) with live steam at 4 bars and 150°C. The resulting thin paste, now concentrated to 18% solids as the result of some condensation of the steam, drops into the batch-making tank containing agitated water and is quickly dispersed. The other ingredients are then added, after which the pH is adjusted to about 8.2 at 25°C.

Starch dispersions have been corrosive to cans, but if ingredients with a low chloride content are selected, and sometimes if a modest amount of corrosion inhibitor is added, a single-lined can is sufficient for a two- or three-year shelf life. One starch formula that uses 0.04% sodium benzoate inhibitor has been successfully marketed in a plain 4.48 g/m^2 (0.20 lb/ft^2) container. If the starch contains a significant amount of chloride ion corrosion promoter left over from sodium hypochlorite (NaClO) bleaching operations, corrosion can become a significant problem and a double-lined can must be used, along with 0.20% sodium nitrite or similar inhibitors. Three typical starch formulations are shown in Table 24.

TABLE 24. SPRAY STARCH FORMULATIONS

Ingredients	Formula A (%)	Formula B (%)	Formula C (%)
Amizo No. 513 Pearl Starch	2.30	----	----
Penford Gum 290 or Equivalent	----	2.75	----
EO-Size 5795 Starch or Equivalent	----	----	3.00
Sodium Tetraborate 10-Hydrate	0.30	0.40	0.45
Silicone emulsion LE-463, 346 or equal. (60% Active Ingredient)	0.40	0.50	0.45
Silicone antifoam emulsion, as SAG-470 by Union Carbide	0.15	0.10	0.10
Sodium Nitrite or Sodium Benzoate	0.15	----	0.10
Fragrance	0.02	0.03	0.03
Glutaraldehyde (50%) or Formaldehyde (37% in Water)	0.04	0.05	0.03
Optical Brightener; as Tinopal 4BM	0.02	----	----
Deionized Water	91.20	91.70	90.00
Isobutane A-31	5.50	6.00	5.84

Note: Adjust pH to 8.4 ± 0.2 at 25°C, using triethanolamine (99%) or a 10% solution of sodium hydroxide.

The optical brightener ingredient is now rarely seen. It adds cost and has only slight marketing appeal.

If available, glutaraldehyde is preferred over formaldehyde, since the latter is less effective as a microbicide and may be a low-order carcinogen. Glutaraldehyde is marketed by Union Carbide Corporation.

The sodium tetraborate, added as the 5- or 10-hydrate, combines chemically with the starch to give it better properties, such as less buildup on the sole plate of the iron, and also functions as a corrosion inhibitor. The silicone emulsions increase the easy slip of the iron over the cloth, so that the user has less fatigue and almost no wrinkling or bunch-up problems. The starch foam should quickly absorb into the textile, and should not be pushed around by the iron. If this is not the case, either a 10% or 100% silicone anti-foamant is added. This ingredient consists of a silicone oil that contains billions of tiny sharp splinters of silica (SiO_2). The silica acts to puncture the foam bubbles so that they quickly collapse. Antifoams are used in about half of all starch products.

Since starch solutions are nutrients for bacteria, yeasts, molds, fungi, and rickettsia, it is necessary to perform the following steps during production:

- Pasteurize the deionized water, or filter it at 0.2μ to remove microorganisms. (Resistant strains of pseudomonads are hard to kill chemically, but they are eliminated by heating for one minute at 40°C or higher.)

- Sanitize the batch-making tank, the concentrate filler, and all the pumps, filters, piping, and other equipment.

- Do not hold the starch concentrate for more than 72 hours, and then only in covered tanks.

- Flush the deionizer beds periodically with a strong formaldehyde solution, to prevent the proliferation of microorganisms on the resins.

Some fillers also run a Total Plate Count (TPC) study on starch batches and finished aerosols. Since these tests require 48 hours for a reliable result, the batch will normally be packed into cans and corrective action will be severely limited. The best result will indicate that there are "fewer than

10 microorganisms per cubic centimeter of product." Therefore, the possibility of proliferation still exists. Experience with starches, fabric
finishes, mousse products, and others that can support microbial growth
suggests that the possibility of growth is low once the aerosol can has been
packed. For example, non-facultative aerobic bacteria generally die from lack
of oxygen. In the rare instances of starch contamination, bacteria have
caused "ropiness," or little tendrils of retrograded material in the product,
leading to valve problems and odors from triethylamine and other substances.

Most starch formulas use 0.03 to 0.035% Formalin (37% formaldehyde, HCHO,
in water). Sodium o.phenylphenate is still occasionally used. However, a
preferred material is glutaraldehyde $[CH_2(CH_2 \cdot CHO)_2]$. This material has very
broad spectrum activity, is low in cost, effective at concentrations even
lower than formaldehyde, and does not sometimes sting the nose of the user
during the ironing process.

A significant amount of work has gone into refining the designs of valves
that will be best for starches. The best spray pattern, for instance, is a
"doughnut" or torus shape, without "hot spots" or areas of extra-heavy product
concentration. This pattern gives the most uniform spray density as the spray
is sprayed across the fabric.

Most starches in the U.S. now use either vertical-acting or toggle-type
valves whose button and stem are separate components. In general, the buttons
are of the two-piece type, with a plastic insert of 0.46 to 0.50 mm orifice
design. However, a new "pseudo-mechanical breakup" one-piece button by the
Precision Valve Corporation is now being used commercially.

During product development, valve candidates are tested against production or commercial control units for the incidence of valve problems. The
test often involves 72 dispensers and lasts two weeks. The caps are left off
to enhance product evaporation at the valve.

Protocols differ, but one starts on a Monday, with 48 cans being sprayed
for 5 to 10 seconds on the following schedule: 2, 2, 3, 2, 2, and 3 days, and

24 cans being sprayed every seven days. The results are recorded in the
following ways:

- Clear (Normal);
- Streams--three seconds or less--then clears;
- Streams--sustained streamer--over three seconds; and
- Plugger.

If more than two sustained streamers or one plugger are encountered, the
valve or formula should be adjusted to be more reliable. If the control,
using a different valve, gave acceptable results, this suggests that the
experimental valve needs improvement, not the formula.

Starch products work better on cottons and on 50% cotton and 50% polyester fabrics than on textiles of higher polyester content, i.e., the so-
called "synthetics." Some marketers have developed Fabric Sizing or Fabric
Finish aerosols especially for these synthetics. As the market for synthetics
diminished during the last 10 years or so, consumers began to discover that
they liked the performance of these fabric finishes on straight cottons and
high-cotton blends. They did not have the relative stiffness of the starches,
but gave the fabrics lubricity, a better feel or handle, and an impression of
heaviness and brighter colors, making them seem more like new garments. since
the product was based on sodium methylcellulose gums, the Fabric Finish
produced a somewhat higher-quality spray and had a higher-quality image than
the starch. It is a useful supplementary product, and sales were estimated to
be about 12 to 14% of the starch market in the U.S. in 1988. Table 25
presents a formulation for aerosol fabric finish.

Heavy-Duty Hard-Surface Cleaners

This product was an outgrowth of the window cleaners. It is sold in two
versions: one with a microbicide, and the other without such an ingredient.
If the cleaner has a microbicide and is labeled accordingly, in the U.S. it
falls under the jurisdiction of the Environmental Protection Agency (EPA)

TABLE 25.   FABRIC FINISH FORMULATIONS

INGREDIENTS	FORMULA (%)
Sodium Methyl Cellulose (Combination of low and medium-low viscosity products). Technical Grade preferred.	0.9
Polyethylene Glycol (Mol.Wt. 400)	0.9
Coco-β-amino-propionic Acid (60% in water)[a]	0.02
Dimethylsilicone Emulsion (60% in water)	0.3
Silicone Anti-foam (100%)	0.03
Ammonium Hydroxide (29% NH_3 in water)[b]	0.02
Glutaraldehyde (50% in water)	0.03
Sodium Nitrite	0.05
Fragrance	0.05
Deionized Water	92.70
Isobutane (A-31)	5.00

[a]Deriphat 151, by Henkel Corporation.

[b]Used to neutralize the Deriphat 151 (Acid) and adjust the pH value to 8.6 ± 0.2 at 25°C.

FIFRA and is subject to a heavy burden of microbiological testing, label review, and confidential formula disclosure before marketing. The delay period is currently one to two years, and an annual registration fee must also be paid to EPA and to state agencies. Some marketers elect to include microbicides in their products and sell them without making any claims, since the cost differential is extremely small.

The first hard-surface cleaner was introduced around 1961. The products are characterized by a combination of non-ionic and tetrasodium ethylene-diaminetetraacetate (Na_4EDTA) detergents, plus various alcohol- or glycol-based solvents in a water base. Two typical hard-surface cleaners are shown in Table 26.

The tetrasodium EDTA, present at 1.90% (Formula A) and 1.52% (Formula B) is effective at removing calcium carbonate, which it does by a sequestering action, producing soluble calcium ethylenediaminetetraacetate $[Ca_2(EDTA)]$. The dissolution process is a slow one if the lime has any significant thickness; therefore, the main thrust is one of preventive maintenance. Rust deposits are similarly dissolved. These uses suggest various bathroom applications.

Higher pH versions, sometimes with 25% deodorized kerosenes added, have been used as whitewall tire cleaners. Anhydrous versions, generally containing about 5% non-ionic surfactant, 20% xylenes, 72% deodorized kerosene, and 3% carbon dioxide are used for cleaning the exterior of car engines. After use, the cleaner can be flushed away with tap water. It is generally advisable to perform the cleaning operation outside on a cool engine that is not running. Hydrocarbon propellants, such as propane (A-108), have been used for these products, but they are not recommended because of their flammability.

A second anhydrous version, in this case not containing any surfactant materials, is carburetor and choke cleaner. It typically contains 60% toluene or (better) xylenes, 30% diacetone alcohol or acetone, and 10% propane. It is obviously extremely flammable, and should be used in small amounts and with care in an open and well-ventilated area. The engine should be cool and

TABLE 26.   HARD SURFACE CLEANER FORMULATIONS

INGREDIENTS	FORMULA A (%)	FORMULA B (%)
Atlas G-3821 Detergent	-	0.50
Tergitol 15-S-9 (Non-ionic surfactant)	0.50	-
Tetrasodium EDTA (38% in water)[a]	5.00	4.00
Triethanolamine - 85%	-	1.00
Propylene Glycol Monobutyl Ether	5.00	6.00
Sodium Meta-Silicate 5-Hydrate	0.10	-
Sodium Sesqui-Carbonate	-	0.10
Morpholine	0.20	0.15
Ammonium Hydroxide (29% NH_3 in Water)	-	1.00
Sodium Hydroxide (50%) or Citric Acid (50%)[b]	q.s.[c]	q.s.[c]
Fragrance	0.10	0.15
Deionized Water	82.10	79.90
Isobutane (A-31)	7.00	7.20

[a]Although a specific surfactant was mentioned, any one or more of the following may be used:

 • Linear primary alcohol polyglycol ether (9 to 12 mol ethylene glycol (ETO); average);
 • Linear secondary alcohol polyglycol ether (9 to 12 mol ETO; average); or
 • Nonylphenol polyoxyethylene (9 to 13 mol ETO; average).

[b] These reagents are used to adjust pH value to 10.5 ± 0.2 at 25°C.

[c] Quantum sufficit (a sufficient quantity).

turned off. Any excess should be removed before the car is restarted. The use of a nonflammable propellant, such as HCFC-22 or carbon dioxide, would act to make the overall product only slightly less flammable.

The final version of a hard-surface cleaner is oven cleaner. There are both caustic formulas and "pre-caustic" formulas. The caustic ones use from 4 to 8% sodium hydroxide activated by triethanolamine to cut through the varnish-like deposits of grease and food spatters on oven surfaces. The other form contains alkali metal acetates and sometimes other organic salts in a water and surfactant slurry. The product is sprayed on the oven surfaces, after which the oven is closed and heated. This causes the organic salts to pyrolyse via a distinctly two-stage process, producing the oxide, carbon dioxide, and water. The oxide then hydrates to the hydroxide form, which begins dissolving the baked-on greases and other residues. Table 27 presents the formulations of oven cleaners.

The caustic nature of Formula A in Table 27 allows it to rapidly corrode aluminum surfaces, and this is often mentioned on labels. Under U.S. regulations, if an oven cleaner contains more than 2% of a caustic such as sodium hydroxide, the dispenser must be fitted with a child-resistant closure. The same regulations permit one exempt package size designed for homes without children and for adults with physical problems like arthritis who would otherwise have great difficulty using the product. In practice, the market-place has shown a strong preference for the products without the special closures, so that dispensers that have this feature are now only a token part of the overall sales picture. Self-cleaning ovens in the U.S., Europe, and other areas will reduce sales of aerosol oven cleaners.

The other type of heavy-duty cleaner is designed for the specialty cleaning of textiles. The best known is the pre-laundry cleaner stain remover, which is sprayed directly onto a stain and onto the inner neck band of shirts, shirt cuffs, and other areas where dirt and grime seem to concentrate. After spraying, the garment may then be laundered. Some formulations contain enzymes for the more effective removal of proteinaceous stains; e.g., grass stains or bloodstains. Others use a mild detergent system and claim

TABLE 27.   OVEN CLEANER FORMULATIONS

INGREDIENTS	FORMULA A (%)	FORMULA B (%)
Potassium Formate	-	6.0
Potassium Acetate	-	6.0
Sodium Hydroxide	5.0	-
Calcium Dodecylbenzene Sulfonate	-	3.0
Compatible Thickener	-	0.5
Sodium Nitrite	0.2	0.2
Triethanolamine - 99%	1.0	-
Tetrasodium EDTA - 38%	1.0	-
Deionized Water	87.8	78.3
Isobutane (A-31)	5.0	6.0

that the treated garments may be stored for several days before washing, if so desired.

The two major formula types are anhydrous and water-based. The latter generally contains 25 to 40% water. Originally, soil removal was accomplished by using 20 to 25% perchloroethylene, in addition to the usual anionic/non-ionic detergent system and hydroxylic solvents. The perchloroethylene ($Cl_2C=CCl_2$) was a major benefit to the cleaning activity, doing such a good job, in fact, that many users complained that their clothes were cleaner and whiter where the product was applied than in the other areas. Some wondered if the product contained a bleaching agent. The marketers maintained a service for taking care of shirts and other garments submitted to them by consumers for correction or replacement, and what they normally did was to immerse the entire item in the concentrate for a few minutes, rinse it off, dry the garment and return it. The super-cleaning ability of the aerosol product had removed soils that resisted ordinary cleaning methods and that had built up on the garment over months of use, turning the cloth slightly grey or slightly tan. Eventually, the perchloroethylene was deleted, after the Bruce Ames "mutated Salmonella" test suggested that it might be a carcinogen and after a number of more odor-conscious consumers complained that traces of the chlorocarbon odor could be detected in clothes even after they were automatically washed, dried, and ironed. The two main formula types of textile cleaners are illustrated in Table 28.

The isopropanol functions as a mild cleaner, but (just as importantly) as a foam destabilizer and suppressant. Ethanol may also be used for this purpose. In either case, percentages may vary according to the foaming tendencies of the overall formula.

Carpet and Rug Cleaner

This unique product was introduced around 1964 by S. C. Johnson & Son, Inc. It is presented in a very large can, such as the 75x192 mm (USA: 300x709) or the new, necked-in 72x261 mm (USA: 211-213/214x1005), which have

TABLE 28.   PRE-LAUNDRY CLEANER FORMULATIONS

INGREDIENTS	FORMULA A	FORMULA B
Linear primary or secondary alcohol polyglycol ether [2 to 4 mol ethylene glycol (ETO)]	12.0	-
Linear primary or secondary alcohol polyglycol ether (7 to 10 mol ETO)[a]	12.0	10.0
Diethylene Glycol Monomethyl Ether	12.0	5.0
Sodium Laurate/Myristate	0.4	-
Isopropanol - 99%	4.0	5.0
Low-odor n.Paraffinic or iso.Paraffinic Solvent (C_{10} - C_{14} Hydrocarbons)	20.0	786.7
Ammonium Hydroxide (28% NH_3 in Water)	0.5	-
Fragrance (Typically lemon/lime)	0.5	0.5
Enzyme Concentrate (Optional)	1.0	-
Deionized Water	30.1	-
Propane A-108 or Propellants A-85	7.5	-
Carbon Dioxide	-	2.8

[a]May be replaced with octyl or nonyl phenol polyoxyethylene (9 to 13 mol ETO) or other non-ionics of similar HLB value.

capacities of about 820 and 980 mL, respectively. This allows one can to clean a carpet of maximum area.

The products use sodium lauryl sulfate, which acts to pull the dirt and grime out of the carpet fibers and then dries so that vacuum-cleaning can effectively remove it. An emulsified polymer is included to prevent rapid re-soiling of the absorbent fibers. Table 29 gives a typical carpet cleaning formula.

Intensive wetting of the warp and woof of the carpet is not desired, as would occur if sodium stearate/palmitate soaps were to be used. Excessive wetting lengthens drying time, and might also cause mold formations at the base of the carpet or rug. The sodium/magnesium lauryl sulfate combination wets only the surface of the fibers, where most of the dirt is collected. In particular, the magnesium lauryl sulfate helps surround the dislodged dirt into a more friable, dried mass on the surface of the fibers, for easy removal with a vacuum cleaner.

The sodium lauryl sarcosinate functions as a corrosion inhibitor, more or less specific to lauryl sulfate ion and ethoxylated or propoxylated lauryl sulfate moieties. However, for it to function well, there must be a virtual absence of chloride ion, bromide ion, and copper ion. The highly purified "toothpaste" grade of sodium lauryl sulfate (SLS) is acetone extracted or otherwise treated to remove any chloride ion that may be present, depending on the method of synthesis used. Sodium nitrite has often been added to these formulas as an additional corrosion inhibitor.

The upholstery shampoo is a related aerosol product that uses such detergents as sodium lauryl sulfate or morpholinium stearate, plus ingredients such as lauryl-monoethanolamide as a corrosion inhibitor and foam stabilizer. The foam is worked into the upholstery covering with a rough cloth or soft-bristle brush, then allowed to dry before removal. A water-wipe is often used to remove the last bits of product, so that a slightly soapy feeling will not be noticed.

TABLE 29. RUG AND CARPET CLEANER PRODUCT FORMULATION

INGREDIENTS	FORMULA (%)
Sodium Lauryl Sulfate (very low in Chloride Ion)[a]	1.60
Magnesium Lauryl Sulfate (very low in Chloride Ion)[b]	1.20
Sodium Lauryl Sarkosinate - (30% in Water)[c]	3.00
Styrene Maleic Anhydride Copolymer - (15% in Water)	20.00
Optical Brightener; as Calcofluor SD (Optional)	0.02
Ammonium Hydroxide (28% NH$_3$ in Water)[d]	0.16
Fragrance	0.08
Deionized Water	66.44
Isobutane A-31	7.50

[a]As Maprofix 563, by the Onyx Division of Witco Chemical Co.

[b]As Maprofix Mg.

[c]As Maprosil 30.

[d]Used to adjust the pH value to 9.8 ± 0.2 at 25°C, although up to about 1.5% may be used if the clean odor of ammonia (NH$_3$) is desired.

Silica-Based Absorbent Fabric Cleaners

A relatively unique aerosol product uses the extreme absorbency of very finely divided silica powder to literally soak up stains by capillary action. Silica, which has been made by the pyrolysis of silicon tetrachloride, is able to absorb hundreds of times its own weight of various liquids, even greases and gels, and this principle is used here. The silica, in the form of an essentially nonflammable slurry in 1,1,1-trichloroethane, is sprayed onto the fabric to be treated. The solvent quickly evaporates, causing the silica powder to absorb any available liquid materials. After complete drying, the loaded silica is brushed off the cleaned fabric, using light strokes, so as not to embed it in the fiber matrix. The aerosol dispenser often comes with a special plastic cap whose top is molded to have 100 to 200 thin, comb-like tines or bristles. 'The cap is used to brush off the silica.

A typical formulation follows:

ABSORBENT SILICA CLEANER FORMULATIONS

Ingredients	Formula (%)
Fumed Silica Powder	6.00
1,1,1-Trichloroethane	68.00
Isopropanol - 99%	10.00
Fragrance.	0.05
Propane A-108	15.95

The selection of silica powder and a valve with optimum design features are keys to success, since with an incorrect combination, valve plugging may occur. The user can correct this problem only 40 to 60% of the time. There are also considerable problems with evaporation, concentrate losses, toxicological response to 1,1,1-trichloroethane vapors (unless used in a well-ventilated room) and weight control in the manufacture of these products, so that one should not undertake their manufacture lightly.

These cans, when actuated under totally non-conductive conditions, will build up a static charge in the 67,000 to 285,000 V range, based on the results of one fairly large study. This does not adversely affect the consumer in any way, but if a filled can is jammed, defective, or otherwise quickly discharges the contents in a gas house while momentarily not grounded, the spark to a nearby grounded surface may cause ignition of the discharge plume, perhaps with serious consequences. No viable corrective methods for this phenomenon have yet been devised.

Air Fresheners

The air freshener was the second aerosol product to be developed commercially, after insecticides. It was marketed in the U.S. as early as 1948, mainly by oil companies, and then by chemical specialties marketers such as the Colgate-Palmolive Company. The formulas were initially combinations of 1% fragrance, 15% low-odor petroleum distillate, and 84% CFC-12/11 (55:45), until about 1961, when the S. C. Johnson & Son, Inc. firm began to market their line of "Glade" Air Fresheners in a water-based form. These formulations now make up the largest segment of this category. The remaining segments are the "super-dry" sprays, typically containing 99% propellants, and the alcoholic types that average about 50% ethanol. Typical examples of the three versions are presented in Table 30. The use of dimethyl ether propellant in Formula B is justified by the increased solvency of perfume resins that might otherwise precipitate.

As mentioned earlier, Formulas B and C have a Volatile Organic Compound (VOC) level of essentially 100 percent. After February 28, 1990, the State of New Jersey (U.S.) has forbidden the marketing of these formulas unless the VOC content is somehow reduced to 50% or less. The use of 1,1,1-trichloroethane (not a VOC, though it has a potential for stratospheric ozone depletion) is not permitted. Ultimately, it may be necessary to use a combination of something like 6 parts water and 44 parts HFC-152a (replacing 50 parts of Propellant Blend A-60) to be in compliance with the regulations. This change will have a major effect on the retail cost of these products.

TABLE 30.   AIR FRESHENER FORMULATIONS

INGREDIENTS	FORMULA A (%)	FORMULA B (%)	FORMULA C (%)
Fragrance	1.00	1.50	2.00
Odorless Petroleum Distillates	6.28	-	6.00
Lanpolamide 5 Liquid (Croda, Inc.) PEG Lanolinamide and PEG Lanolate ester - 50% in Deodorized Kerosene (HLB - 3.65)	0.72	-	-
S.D. Alcohol 40-2 (Anhydrous)[a]	-	-	38.00
Sodium Benzoate	0.15	-	-
Deionized Water	59.85	-	4.00
Propellant Blend A-60[b]	32.00	90.00	50.00
Dimethyl Ether	-	8.50	-

[a]Specially Denatured ethanol, where 400 g of tertiary butanol $[(CH_3)_3COH]$ and 42 g of brucine sulfate are added to every 3,600 liters of anhydrous ethanol.

[b]Contains typically 62 weight percent isobutane, 2 weight percent of n.butane, and 40 weight percent of propane.

Disinfectant/Deodorant Sprays

This category has often been compared with air fresheners, but there are more differences than similarities. First, the "D/D" products are regulated by the U.S. EPA, so that planning and formula development should be carried out at least three years before the marketing phase begins. Secondly, most of the label is given to a description of the formula, disinfectant claims, and directions for disinfecting hard surfaces. The ability of the product to function as a space spray is limited by the low levels of propellant used, since the main use is as a surface spray, and labeled uses limit space spraying to storage rooms, closets, and other enclosed spaces for deodorizing purposes only. (Fragrance benefits are not mentioned on the label, although a pleasant fragrance is always included, even in "Hospital Strength" D/D products.)

Two formulation types and two propellant types are currently in use. The base product contains either an o.phenyl-phenol system or a quaternary ammonium disinfectant system in a hydro-alcoholic solution. Either 5% carbon dioxide or about 20% hydrocarbon propellant blend is used as the pressurizing medium. In terms of units sold, the o.phenyl-phenol and carbon dioxide system is probably the most popular.

The EPA requires that the labels of these products list the active ingredients, plus certain other data. An example from the label of one such product is shown below:

Active Ingredients:

n-Alkyl (60% C_{14}, 30% C_{16}, 5% C_{12}, 5% C_{18}) dimethyl benzyl ammonium chlorides................................0.072%

n-Alkyl (68% C_{12}, 32% C_{14}) dimethyl ethylbenzyl ammonium chlorides......................................0.072%

Ethanol...53.088%

n-Alkyl (92% C_{18}, 8% C_{16}) n-ethyl morpholinium ethyl sulfate...0.040%

Inert Ingredients:

46.728%

Contains sodium nitrite

The first two ingredients are available as BTC 2125M, which is sold as a 50% active ingredients solution (and in other strengths) by the Onyx Division of the Witco Chemical Company. Similarly, the last n-Alkyl compound is sold as Atlas G-271, generally as a 35% active ingredient solution, by ICI America, Inc. Two examples of these formulas are provided in Table 31.

The quaternary ammonium chloride products came along well after the market for D/D aerosols was well established and approaching 100,000,000 units sold a year in the U.S. The strengths and weaknesses of their antimicrobial spectrum of efficacy is different from that of o.phenyl-phenol and its close derivatives, as would be expected. Also, since the quaternaries are much less volatile than the substituted phenols, the protective effects may last longer. This may be important when considering regrowth potential for molds in leather, wood, books, and other relatively porous substrates. No one has attempted to market a product containing both microbial types, perhaps because of the degree of toxicological and microbiological testing that would be required.

Since they have chloride ion (a strong corrosion promoter) double-lined cans and heavy amounts of strong corrosion inhibitors have been required to achieve an adequate shelf life for the quaternary ammonium formulations. For some time, combinations of sodium nitrite and morpholine were preferred for the inhibitor system, but after it was found that up to about 10 parts per million of morpholinium-N-nitrosamine (a carcinogen) could be formed in situ over one year of room-temperature storage, marketers acted to change the sodium nitrite to sodium benzoate and eliminate the reaction.

Disinfectant Cleaners

This type of product was partially covered under "Hard Surface Cleaners" (see Table 26), but the disinfectant version adds a new dimension of cleaning that is generally appreciated by the consumer. Most of the larger marketers of heavy-duty cleaners are able to cope with EPA's requirements for pre-marketing registration, plus federal and state fees, and have preferred this type of presentation. The disinfectant cleaner is really nothing more than

TABLE 31.   DISINFECTANT/DEODORANT FORMULATIONS

INGREDIENTS	FORMULA A (%)	FORMULA B (%)
o.Phenyl-phenol (98% purity)	0.110	-
BTC-2125M (50% in water)[a]	-	0.288
Atlas G-271 (35% in water)[a]	-	0.114
S.D. Alcohol 40-2 (Anhydrous)[b]	73.380	52.068
Fragrance	0.110	0.110
Sodium Benzoate	0.200	0.220
Morpholine	0.200	0.200
Deionized Water	21.000	25.000
Propellant Blend A-40[c]	-	22.000
Carbon Dioxide	5.000	-

[a]For chemical compositions, see preceding page.

[b]For chemical composition, see note 'a' of Table 30.

[c]10 wt % propane and 90 wt % isobutane.

the standard type, except for the inclusion of 0.20% or so of biocidal material in the formula. When a quaternary microbicide is used, the formula has to be adjusted to eliminate incompatible anionic surfactants that might precipitate the active cationic moiety. Some remain acceptable, as will be seen in the formulation presented in Table 32.

Normally, two types of valves are used for both these and the regular hard surface (basin, bath, and tile) cleaners. The can may be used in different positions, including some where the dip tube may protrude into the gas space. The simplest and least costly approach is to use a valve with a very large diameter, a "jumbo" dip tube, with an inside diameter of about 6.4 mm. For the relatively long cans in general use, such tubes will contain 7 to 8 grams of product. If the container is turned upside down--for instance, to more comfortably spray the base of a toilet bowl--the special dip tube will hold sufficient product for about 6 seconds of spray time. After this, gas will be emitted, signalling the consumer to reverse the can for a second or two. In the second approach one might use the Sequist Valve Company Model NS-36 (Ball-check) valve. A 4-mm diameter stainless steel ball travels in a short plastic slot, just below the valve. With the can upright, an orifice at the bottom of the slot is closed off, forcing the product to travel up the dip tube and through the valve. With the can inverted, the ball closes off an orifice at the opposite end of the slot. This acts to plug the opening from dip tube to valve and at the same time opens a "vapor-tap" type orifice directly into the valve chamber. The valve has only two minor deficiencies: it always leaks slightly between the plastic and the ball, to give a vapor-tap effect, and secondly, it works poorly when the can is in an essentially flat position. The price is significantly higher than that of the standard valve or jumbo dip tube valve.

A good delivery rate for the hard-surface cleaners is about 1.23 g/sec at 21°C, at the 460-mm vacuum crimp pressure of about 2.54 bar at that temperature. A valve with a 0.46-mm stem and 0.41-mm MB-ST button will provide the desired rate.

TABLE 32.   DISINFECTANT CLEANER FORMULATIONS

INGREDIENTS	FORMULA (%)
Sodium Meta-Silicate 5-Hydrate	0.10
Tetrasodium EDTA (38% A.I. in Water)[a]	4.12
BTC 2125M (50% A.I. in Water)[b]	0.40
Sodium Benzoate	0.10
Sodium Tetraborate 10-Hydrate	0.10
Morpholine	0.20
Ammonium Hydroxide (As 29% NH_3 in Water)	1.10
Atlas G-3821 Non-ionic Surfactant[c]	0.50
Butyl Cellosolve (or similar)[d]	6.00
Potassium Hydroxide (45% A.I. in Water)	0.05
Fragrance	0.15
Deionized Water	80.18
Isobutane A-31	7.00

[a]Tetrasodium Ethylenediamine-tetraacetate, such as Cheelox BF-13, or Versene 30 (Dow).

[b]See previous pages for complex formula of ingredients.

[c]By ICI America, Inc.

[d]By Union Carbide Corporation. Propylene Glycol Monomethyl Ether may also be used.

Paint Products

    In 1988, the U.S. paints and coatings industry marketed approximately
325,000,000 units, ranging from very small touch-up paints to large-size units
for domestic or industrial furniture finishing. A substantial number of
filling plants specialize in self-fill or contract filling operations. It is
a complex area, with five main categories: enamels, lacquers, varnishes,
stains, and primers, with subgroups of each. Large numbers of colors have
also to be considered. The largest sales are for the alkyd- and acrylic-base
paints. Both are available in anhydrous and water-based formulations,
although the water-based techniques are better developed in some countries
than others, as is the use of dimethyl ether as a paint propellant. The
formulas to follow illustrate a bronze metallic specialty lacquer (anhydrous),
two alkyd types and an acrylic type. The last three are based on some
excellent work by DuPont that has been widely distributed. The term lacquer
refers to a coating that dries by the simple evaporation of the solvent
system. Originally, it related to the cellulosic varieties, but these have
been almost completely displaced by the thermoplastic acrylics. The acrylics
have better resistance to mild chemicals, weather, and sunshine. They are a
preferred base for various metallic finishes (aluminum, bronze, and gold
powder finishes) because of their low acid number and water-white color. Four
different paint formulations are illustrated in Table 33.

    A prototype valve that might be evaluated is the Newman-Green Model R-10-
123 (0.33-mm vapor-tap), but with a butyl rubber seal gasket. The actuator is
a No. 120-20-18. This valve delivers the four products shown in Table 33 at
about 0.95 g/sec at 21.1°C.

    During the development of various paint aerosols, alterations in the
formula or valve may be required if the applied product exhibits low gloss,
blushing, sagging, bubbling, peeling, deleafing of metallics, valve plugging,
poor adhesion, low durability, or other problems. For example, adding more
xylenes to Formula A would slow down the final drying of the film, resulting
in better smoothness and higher durability. The disadvantage must be weighed
against the two advantages, keeping in mind that the consumer will note the

TABLE 33. VARIOUS AEROSOL PAINT FORMULATIONS

INGREDIENTS	ACRYLIC METALLIC FORMULA A (%)	ACRYLIC FORMULA B (%)	ALKYD FORMULA C (%)	ALKYD FORMULA D (%)
Acryloid B72 (50% A.I.)	8.0	-	-	-
Acryloid A101 (40% A.I.)	1.0	-	-	-
Carboset 514H (40% A.I.)	-	25.00	-	-
Gold Powder #6238	4.0	-	-	-
Tint Ayd (Black WD-2350)	-	5.00	5.00	5.00
Beckosol 13-400 (75% A.I.)	-	-	13.00	13.00
Ammonium Hydroxide (29% NH_3)	-	-	1.15	1.15
Titanium Dioxide Powder, R-940	-	1.00	2.00	-
Propylene Glycol Monomethyl Ether	2.0	5.00	5.00	5.00
Isopropanol	-	8.00	8.00	8.00
Nonylphenoxy Polyethoxy Ethanol	0.1	0.35	0.45	0.50
Fluoroacrylic FC-430 Surfactant	-	0.02	0.02	0.02
Hi-Sil T-600 (Silica)	-	0.14	0.14	0.14
Magnesium Aluminum Silicate	-	0.30	0.12	0.15
Drier: Cobalt Hydro Cure II	-	-	0.10	-
Drier: Zirconium Hydro Cem	-	-	0.08	-
Toluene	28.2	-	-	-
Xylenes	12.4	-	-	-
Acetone	15.0	-	-	-
Deionized Water	-	10.19	19.94	22.04
Hydrocarbon Propellant Blend A-85	28.9	-	-	-
Dimethyl Ether	-	45.00	45.00	45.00
Pressure (460-mm Vacuum Crimp bars at 21.1°C.)	3.4	3.9	3.8	3.8

(Continued)

TABLE 33.   (Continued)

Ingredient Sources:

Acryloids	Rohm & Haas Company
Carboset	B. F. Goodrich Company
Gold Powder	U.S. Bronze, et al.
Tint Ayd	Daniel Products Company
Beckosol	Reichhold Chemical Company
Titanium Dioxide	E. I. duPont de Nemours & Co., Inc.
Propylene Glycol MM Et	UCAR PM - Union Carbide Corporation
Nonylphenoxy Poly.	Triton N-401 - Rohm & Haas Company
Fluoroacrylic	3M Company
Hi-Sil	PPG Industries, Inc.
Magnesium A.S.	Attagel 40 - Engelhard Corporation
Driers	Mooney Chemicals, Inc.
Dimethyl Ether	Dymel A - E. I. duPont de Nemours & Co., Inc.

disadvantage rather soon after the product is used, but may not detect the other differences until later, if at all.

Paints and coatings are generally packed with a small glass marble that helps agitate settled material back into a uniform dispersion. Also, to prevent premature use by children or others, a tamper-resistant and tamper-evident valve cover or protective cap is used. The outer cap is often colored the same as the product within the can, or it may carry a self-adhesive top label to help the customer make selections.

Furniture Polishes

The original aerosol furniture polishes were introduced around 1950. They contained self-polishing floor waxes in a simple oil-in-water emulsion form. In 1955, silicone emulsions were included, since they added lubricity and made the rubbing out process much easier. They also improved the sheen and conferred water resistance to the polish. At first, formulators added an intermediate viscosity silicone, such as Dow-Corning DC-200 dimethylsilicone fluid (1,000 cstks), at a low-volatiles level about the same as that of wax: 0.7 to 1.5% of the total. But as they found that silicones soaked into the polishing cloth more readily than wax, they began to increase silicone levels. In addition, it was found that combinations of lower- and higher-viscosity silicones functioned better than the single intermediate viscosity type. The higher-viscosity silicone added shine or brilliance, but too much caused the polished surface to be subject to marking. Two illustrative examples are shown in Table 34.

The preparation of furniture polish concentrates can present fire and explosion hazards, especially if the more volatile aliphatic hydrocarbons are used, such as Isopar C, which has a flashpoint of 5°C. Heating batches of Isopar C to 80°C or so to facilitate the dissolution of waxes has caused four major explosions and subsequent fires. This is because very heavy vapors of the hydrocarbon seep over the tank rim, fall to the floor, and spread outward until a spark or fire source is contacted. Less than 1 volume percent of

TABLE 34.   FURNITURE POLISH FORMULATIONS

INGREDIENTS	FORMULA A[a] (%)	FORMULA B[b] (%)
Wax S and Wax N (1:1 ratio) Hoechst	1.25	1.25
Silicone Emulsion LE-461 (50% A.I.) UCC	1.40	1.40.
Silicone Emulsion LE-462 (50% A.I.) UCC	0.35	0.35
Arlacel C (Non-ionic surfactant) ICI Am.	0.15	1.25
Isopar C or E (C_7 or C_8 iso.paraffinics) Exxon Oil Company	2.00	33.00[c]
Lemon Oil, Technical Grade	0.75	0.60
Glutaraldehyde (50% A.I.) UCC	0.05	0.03
Sodium Nitrite	0.05	0.05
Deionized Water	86.00	44.67
Isobutane A-31	7.00	17.50

[a]Oil-in-water version.

[b]Water-in-oil version. (Better product; more costly.)

[c]Any n-paraffinic, iso.paraffinic, or multi-brancheate low-odor hydrocarbon may be used, at 12 to 36%. About 20% is an average.

flammable vapor in air is required for ignition. Air-tight compounding tanks and good ventilation is required.

A related product is the wood paneling cleaner, conditioner, and polish. Pre-finished plywood wall panels, natural wood kitchen cabinets, and similar surfaces have relatively thin varnished or lacquered surfaces compared with furniture, so that the use of water-based polishes like those just described would result in some water penetration of the wood, and the finish would be gradually lifted or peeled. As a result, these products are anhydrous and de-emphasize the use of wax-type ingredients. The formulation in Table 35 provides good gloss, sealing, and detergent resistance.

Car Windshield De-Icers

The windshield de-icer spray is a product representative of dozens of automotive aerosols. The most effective de-icer is methanol (CH_3OH), and it is used to some extent, despite its well-known toxicity and the corresponding need for special labeling under various U.S. government regulations, such as the CPSC regulations. Isopropanol [$(CH_3)_2CHOH$] and n.propanol ($CH_3-CH_2-CH_2OH$) are less hazardous but are less effective and more costly. Since a simple alcohol or alcohol/water de-icer would allow refreezing of the liquid to occur as soon as the alcohol was sufficiently diluted or evaporated, it is customary to add a certain amount of glycol to formulas. Here again, ethylene glycol ($HO-CH_2-CH_2-OH$) is the most effective, but it also is quite poisonous, so propylene glycol [$HO-C(CH_3)H-CH_2-OH$] is used instead.

If very thin ice films are dissolved by an anhydrous alcohol/glycol product, after which the alcohol largely evaporates, vision will be obscured by the heavy glycol layer that remains. To resolve this final problem, certain amounts of water are included in the formulas. The higher-quality products will have about 20%, while the economy types may have as much as 50 percent. Table 36 presents a typical formulation.

TABLE 35.   WOOD PANEL POLISH FORMULATIONS

INGREDIENTS	FORMULA (%)
D.C. 536 Fluid (An aminofunctional polydimethylsiloxane copolymer - Dow Corning Corporation)	2.00
D.C. 200 Fluid (12,500 cstks) (Dimethylsiloxane polymer - Dow Corning Corporation)	2.00
Witcamide 511 - Witco Chemical Company	1.00
Isopar L and/or Isopar M - Exxon Company	26.50
Isopar K - Exxon Company	65.20
Fragrance	0.05
Isopropanol (anhydrous)	0.25
Carbon Dioxide	3.00
Pressure (460-mm Vacuum Crimp) bar at 21.1°C	7.40

TABLE 36.   WINDSHIELD DE-ICER FORMULATIONS

INGREDIENTS	FORMULA (%)
Methanol - Technical Grade	54.00
Propylene Glycol - Technical Grade	18.00
Deionized Water	25.00
Morpholine	0.10
Span 80 or Igepal CO-410 Non-ionics	0.05
Sodium Benzoate	0.05
Carbon Dioxide	2.80

The Igepal CO-410 (Rohm & Haas Co.) surface active agent is present in the formula in Table 36 because it improves the wetting activity of the formula, allowing it to penetrate more effectively into fissures and cracks in the ice, and then between the ice and the glass, for faster removal.

PESTICIDE AEROSOL PRODUCTS

Pesticides consist of insecticides, insect repellents, disinfectants, rodenticides, nematocides, herbicides, and a host of other products designed to reduce or eliminate pests in size ranges extending from viruses to rats. All these products fall under the purview of the EPA FIFRA if they are made or marketed in the U.S. Other nations have similar regulations. When pesticide products are designed for use on the skin in the form of "outdoor lotions" that protect against solar radiation, poisonous plants, infections from scratches, and also contain insect repellant, the EPA still has control but may consult with other agencies, such as the FDA in this case, before giving pre-market clearance. Information has been presented earlier on the disinfectant cleaner and disinfectant/deodorant spray, which are regulated by the EPA in the U.S.

Insecticides

The insecticide was the first commercial aerosol product, used as early as 1944 for both military and domestic applications. These early sprays were true "aerosols" (unlike any of today's products, except one type) and used 85 to 90% of CFC-12 to disperse the pyrethrin-containing concentrates. The first major segmentation of this product form came in 1953, with the introduction of the bug killer: a coarse spray consisting of at least 75% kerosene-based concentrate, used for surface wetting, instead of the usual space spray format. By 1961, water-based space sprays came onto the market, and many years later this technology was applied to the surface spray as well. Also in the early 1960s, a "whole-house insecticide," or "total release indoor fogger" spray was developed, typically using 85% CFC propellants. Other specialty insect sprays were developed later in the 1960s. They included the wasp and hornet spray, pressurized with nitrogen or carbon dioxide, and which could throw a stream or streaming spray up to 6 meters. A number of pet-stock sprays were also introduced. Later, hormonal flea-control sprays, biocidal sprays, and other types were introduced.

The space sprays are now essentially all water-based, since the other formulations were too costly and could not compete with the obvious economies offered by combinations of approximately 65% water and 30% hydrocarbon propellant. The only exceptions are the total release indoor fogger (TRIF) and toxicant/propellant (T/P) sprays.

The water-based space sprays can be closely compared with the air freshener shown in Table 30, Formula A. By removing the perfume ingredient and replacing it with a toxicant blend, the transition is complete. The water-based space sprays include the flying insect spray, house and garden spray, patio fogger, and a portion of the TRIF products. As a unit, they make up approximately 55% of the insecticide aerosol volume.

The TRIF spray made a difficult transition during 1978, when CFC propellants were banned in the U.S. Since it is designed to be latched open and to discharge the entire contents of the can within two or three minutes, there is a greater inhalation and flammability hazard than is the case with most aerosols, which release only a few grams at a time. The flammability aspect related to two factors: the size of the container (and the number used at one time), and the degree of product flammability. When problems have occurred, they have been caused by gross consumer misuse; for example, when two or more large cans have been set off in a relatively small area containing an ignition source such as the pilot light of a stove (range and oven), gas-fired refrigerator or gas-fired hot water heater. Table 37 presents three forms of commercial formulations for these products.

The relative flammability of the TRIF sprays can be assessed by using a slight modification of the Department of Transportation (DOT) Closed Drum Test in the U.S. The 200-liter drum is laid on its side, with the open end closed off with a film of plastic. A candle is lit at the bottom and the spray is immediately introduced, using the test formula but a different valve more

TABLE 37.   TOTAL RELEASE INSECT FOGGER FORMULATIONS

INGREDIENTS	FORMULA A (%)	FORMULA B (%)	FORMULA C (%)
Pyrethrum Extract - 20%	2.00	2.00	-
Piperonyl Butoxide; Technical	1.00	1.00	-
Emulsifiable Concentrate	-	-	8.00
Petroleum Distillates	12.00	12.00	7.00
Methylene Chloride	-	15.00	-
1,1,1-Trichloroethane	55.00	40.00	-
Deionized Water	-	-	50.00
Propane A-108	30.00	-	-
Isobutane A-31	-	15.00	35.00
HCFC-22	-	15.00	-

compatible with the test procedure than the "latch open" type. The number of grams of product sprayed into the drum until the Lower Explosive Limit (LEL) "poof" is reached and recorded. From that figure, the number of cubic meters that the dispenser can bring to the LEL composition is readily calculated.

Insect Repellents

The usual insect repellent is used to keep users from being bitten or stung by various flying insects. The most common ingredient is N,N-Diethyl-m-toluamide in concentrations of 15 to 30% of the total formula. Sometimes other repellents are added for protection against insects only partially repelled by the DEET active ingredient. They include MGK Repellent 11 and MGK Repellent 264 and are offered by the McLaughlin, Gormley & King Company, of Minneapolis, MN (U.S.). A typical formula is shown in Table 38.

Variations on Formula Type 26 include replacing the hydrocarbon propellant with 4.5% carbon dioxide, replacing the ethanol with isopropanol, and removing the three MGK products, while increasing the level of DEET repellent to about 30 percent.

The transfer efficiency from dispenser to skin or clothing is only about 55 to 65%, making other forms more attractive by comparison. Lotions and sticks are available, as well as roll-on forms.

PHARMACEUTICAL PRODUCTS

These products are generally perceived as those that are inhaled, injected, or otherwise inserted into the body to mitigate or control medical problems such as migraine headaches, asthma, hemorrhoids, etc., or to provide a contraceptive function, such as vaginal contraceptive foam. A few of these products have already been covered in the foams area of this chapter. The primary one that remains is the metered dose inhalant drug (MDID), which represents a U.S. market conservatively estimated at well over 100,000,000 units per year and served by at least 28 brand-named products. As is common with the rest of the aerosol industry, products are self-filled and also

TABLE 38.   INSECT REPELLENT FORMULATIONS

INGREDIENTS	FORMULA (%)
N,N-Diethyl-m-toluamide (95% A.I. min.)	20.0
MGK Repellent 11	2.0
MGK Repellent 326	1.5
MGK 264	1.5
S.D. Alcohol 40-2 (Anhydrous)	54.9
Fragrance	0.1
Propellant A-46 16 wt % propane and 84 wt % isobutane	20.0

TABLE 39.   BETA-ADRENERGIC BRONCHODILATOR FORMULA

Ingredients	g/10.5 g Can	Percentage (w/w)
Terbutaline Sulfate (Drug)	0.075	0.714
Sorbitan Trioleate (Excipient)	0.105	1.000
CFC-11	2.580	24.571
CFC-114	2.580	24.571
CFC-12	5.160	49.144

contract filled. One or more self-fillers also contract fill for their competitors. At this time, all of these products use one or more of the following propellants: CFC-11, CFC-12, and CFC-114. The volume of propellants used is approximately 1,900 kilotonnes (4,200,000 pounds) in the U.S. alone. Table 39 shows a typical published formulation. Others are suggested in U.S. Patent literature and other documents.

Use of hydrocarbon propellants for some of these products is not satisfactory because of production problems related to flammability, the oily, stinging taste they have when inhaled nasally or orally, and their very low density (considered from the standpoint of drug precipitation rates during use).

CFC-11 is slurried with the drug and one or more excipient materials, and this mixture is added to aluminum aerosol cans or bottles. They are then fitted with a ferrule-type meter-spray valve which is hermetically sealed to the container by a clinching or under-tucking operation. The CFC-12, sometimes mixed with CFC-114, is then introduced backwards through the valve.

CFC-11, with a boiling point of about 23°C, is unmatched by any other nonflammable solvent of acceptable toxicology. Its replacement will necessarily depend on the availability of one or more of the "future alternative" HCFC and HFC propellants due to come on the market in 1992 or 1993.

For the 90% of MDID products that use very finely divided microcrystalline drug particles (averaging from 3 to 5 microns), it is important to have a system of low solvency. Otherwise, the larger particles will get still larger and the smaller ones (because of their higher surface energy) will get smaller until they vanish. This disturbance will severely limit the product effectiveness. Even with the optimum particle size distribution, the body's defenses are such that only 7 to 12% of the drug reaches the target areas. With the formation of larger particles in the container, this could drop to below one percent.

The time frame needed for additional toxicological testing of the HFC and HCFC propellants, such as that being done in the Program for Alternative Fluorocarbon Toxicity Testing (PAFTT)I, PAFTT II, and PAFTT III consortium tests sponsored by the chemical producers, is one element of the new product development. Another is actual formula and package development and specifications writing. A third is the opening of each company's "New Drug Application" to the FDA, requesting an "Amended New Drug Application" (ANDA). The entire documentation is reviewed in such procedures, which typically take from 3 to 5 years to complete if there are no problems. One industry concern is that the FDA may not have sufficient staff to process approximately 27 concurrent ANDAs with anything like their usual timing. These considerations suggest that it would be to the advantage of chemical producers to cooperate in their efforts to have new products cleared by the FDA within the generally planned transition period ending about 2000.

The viability of new formulas depends on their solvency and toxicology. Preliminary results from PAFTT will be released in September 1989. Because of the uncertainty about HCFC-123, three possible formulas are suggested here for consideration (see Table 40).

The use of HCFC-124 is optional, since it merely serves to reduce the pressure slightly. The CFC-113 is used as an additive to the slightly flammable HCFC-141b to create a nonflammable blend for slurrying purposes. If the pharmaceutical firms and their fillers can handle a slightly flammable slurrying agent (pure HCFC-141b), there will be no need to use the CFC-113 (or CFC-11).

INDUSTRIAL AEROSOL PRODUCTS

There are numerous aerosols used only in industrial or institutional applications. Two will be considered here: a lubricant spray for pharmaceutical pill- and tablet-making rotary molding machines, and an industrial

TABLE 40.   METERED-DOSE INHALANT DRUG FORMULATIONS

INGREDIENTS	FORMULA A (%)	FORMULA B (%)	FORMULA C (%)
Drug (as a microcrystalline suspension)	0.5	0.5	0.5
Excipient(s)	1.0	1.0	1.0
HCFC-123	13.5	-	-
CFC-113 (or CFC-11)	-	4.5	-
HCFC-141b	-	9.0	13.5
HFC-134a	75.0 - 85.0	75.0 - 85.0	75.0 - 85.0
HCFC-124	10.0 - none	10.0 - none	10.0 - none

adhesive. For the first application, the products must be nonflammable, and leave only a Food Grade [Generally Recognized as Safe (GRAS)-Listed] residue on surfaces to be contacted by the pharmaceutical pill or tablet. A current formulation is shown below:

Rotary Tablet Machine Die Lubricant

Ingredients	Formula (%)
Lecithin (Soy Bean source)	2.0
Sorbitan Trioleate	0.5
Ethanol (Anhydrous)	2.5
CFC-113 (Especially purified)	70.0
CFC-12	25.0

An intermediate step could replace the CFC-12 with a mixture of 10 parts HCFC-142b and 20 parts HCFC-22, reducing the CFC-113 to 65 parts in the process. This would reduce the CFC content by 32 percent.

When the future alternative propellants become available, the formulations shown in Table 41 could be considered. Substantial testing of these prototype formulas in Table 41 would be required as a prerequisite to commercial use.

Adhesive Spray

A typical industrial product is the adhesive used to coat automotive gaskets before setting them in place on engine blocks or other equipment. Aerosol products have a substantial niche in this market area. A typical formulation is illustrated in Table 42.

The product is sprayed onto the gasket while it lies on a waxed paper or other suitable substrate. After a minute or so, much of the methylene chloride will have evaporated, bringing out the stickiness of the resins. After another five minutes, the gasket is ready to be applied to the engine block or other item.

TABLE 41. ROTARY TABLET MACHINE DIE LUBRICANT FORMULATIONS

INGREDIENTS	FORMULA A (%)	FORMULA B[a] (%)
Lecithin (Soy Bean source)	2.0	2.0
Sorbitan Trioleate	0.5	0.5
Ethanol (Anhydrous)	2.5	2.5
HCFC-123	77.0	-
HCFC-141b	-	55.0
HCFC-124	-	30.0
HCFC-22	18.0	10.0

[a]Formula B could replace Formula A if HCFC-123 does not become commercially available.

TABLE 42. GASKET ADHESIVE FORMULATION

INGREDIENTS	FORMULA (%)
Isopropanol	10
Resin 80-1211[a]	5
Stabilite Ester Number 3[b]	5
Methylene Chloride	50
Xylenes	10
Propellant Blend A-70	20

[a]Made by the National Starch and Chemical Company.
[b]Made by Hercules, Inc.

Section II
Alternative Aerosol Dispensing Systems

1. Introduction

An imposing number of packaging alternatives to the standard aerosol
dispenser are available. Several use aerosol containers, but segregate the
propellant gas, and employ a finger-pump, trigger-pump, hand-operated piston
action, a metal spring, screw device, or other mechanism to dispense the
product or form the propellant gas within the container as required. Others
take the form of rather specialized, non-aerosol containers designed to enable
the user to create air pressure or product pressure, or to operate screw-on
finger-pump or trigger-pump metering valves. The pump-sprays, in all their
diverse forms, represent the most widely used alternative. Such packaging
options as stick applicators, pads, etc. offer alternatives to the aerosol
system but do not provide sprays; these will only be briefly described.

A substantial number of aerosol alternatives will be described in Part II
of this report, beginning with those that are most similar to conventional
aerosols--and that may even be considered aerosols by various persons and
authorities--and continuing with alternate packaging forms that bear no
resemblance to aerosol products.

Definitions

The term "aerosol" was used by the scientific community at least as far
back as 1838 to describe dispersions of liquids in a gaseous medium, such as
fog, mists, and clouds, where the particles were true colloids, having
diameters of approximately 0.005 to 0.200 microns (μ). Particles of this
magnitude were able to remain air-borne indefinitely. The smallest particles
are the same size as many larger molecules, such as starches, proteins, and
rubbers, and this part of the definition has not changed over the years. But

the high end of the size range originally defined as the limit of microscopic visibility, has changed greatly. The so-called "coarse aerosol" (to the physicist) now includes dispersions of particles ranging from 0.2 to about 20 microns (μ). Since the particle size distribution of commercial aerosol sprays is generally in the 1 to 100 micron (μ) range, at least some of the sprays meet the expanded classical definition.

Some early definitions of aerosol products were based on particle size. For example, around 1949, the U.S. Department of Agriculture (USDA) designated that the "true aerosol" insecticide was one in which at least 80% of the particles had a mean diameter of 30 microns (μ) or less, and in which no particle could have a mean diameter greater than 50 microns (μ). To meet these requirements, chemists had to design formulas with 80 to 85% or more of propellant. The rationale was that the spray particles had to be very small to remain airborne for two minutes to two hours to control flying insects. The products soon became known as space sprays.

At about the same time, the USDA introduced the "pressurized spray" concept for insecticides that were slightly more coarse. The mass median diameter of all particles had to be about 25μ, and some could be above 50μ. Because the larger particles fell to the floor in less than one minute, marketers had to use label directions that advised the user to spray an additional 25-50% more product into the air space of rooms.

Finally, about 1951, the "residual spray" insecticide was defined. Essentially all particles had to be larger than 50μ, so that such toxicants as Chlordane, Strobane and DDVP (dichlorphos) would not be inhaled to any significant extent. These were used only for spraying baseboards, doorway sills, wasp-nests, and other inanimate surfaces.

The piston-pump insecticide sprayer could dispense dispersions of particles about 25μ in size with deodorized kerosene formulations and those 20μ in size with the more flammable ethanol and isopropanol compositions. The finger-spray and (later) trigger-spray insecticides generally provided distributions of particles in the 30 to 80μ range. Consequently, much more had to be used for the control of flying insects, and the range of action was also much less than that of aerosol dispensers.

The confusion between "aerosol" (the colloid sol) and "aerosol" (the dispenser) has existed since the aerosol industry was born in 1943. In an internal report, the Academic Press Inc. publishing house mentioned that more copies of one of their new textbooks (Aerosol Science, C.N. Davies - Editor, 1966) had gone to recipients in the aerosol packaging industry than to the intended audience of physicists, physical chemists, and meteorologists, mainly because of the lack of contents identification in some advertising and promotional materials. The industry made an attempt to rename itself as the "Self-pressurized Dispenser Division" of the Chemical Specialties Manufacturers Association, Inc. (CSMA) but the proposal was defeated. Today, the words "aerosol" and "self-pressurized" product are used interchangeably.

For the purposes of interstate transportation, the U.S. Interstate Commerce Commission (ICC), now a branch of the Department of Transportation (DOT), defined the aerosol package in 1948 as follows:

"A sealed package containing base product ingredients, in which one or more propellants is dissolved or dispersed, and fitted with a dispensing valve."

Despite the fact that many self-pressurized products thought of as aerosols do not strictly meet this definition, nearly all are currently shipped under Section ORM-D of the tariff. (The definition has been modified slightly over the years.)

Other definitions are listed below without special comment:

CSMA Definition: "A pressurized sealed container with liquified or compressed gases so that the product is self-dispensing."

FDA Definition: "A package consisting of a container and valve, into which is added a base product and propellant, causing the dispenser to be under pressure, and able to discharge the product as a spray, foam, liquid, gel, or other form."

H.R. Shepherd (Book) Definition, 1960: "A container whose contents are expelled through an opened valve by means of the internal pressure of the materials contained therein."

P.A. Sanders (Book) Definition, 1979: (Also used by CSMA) "A self-contained sprayable product in which the expelling force is supplied by a liquified gas."

National Paints and Coatings Association (NPCA) Definition: "A self-contained package which contains the product and the propellant necessary for the expulsion of the former."

British Aerosol Manufacturers Association, Ltd. (BAMA), 1971: "As integral ready-to-use package incorporating a valve and product which is dispensed by prestored pressure in a controlled manner when the valve is operated."

Most of these definitions were created by one person, then approved by a committee or by a brief committee action. Some are ill-conceived or outdated and either do not cover all aerosols, or cover products not commonly denoted as aerosols. In a recent inquiry to the DOT, a product consisting of a mixture of Halon-1301/1211 (20:80) was finally judged to be a non-aerosol and denied the standard aerosol ORM-D exemptions because it contained no base product ingredients. Two materials other than propellant had to be present to be designated an aerosol. The prospective marketer finally added a drop of kerosene (a mixture of ingredients) to 13 Av.oz. (369 g) of the Halon blend, and is now selling the product.

At a recent industry meeting, representatives from the Metal Box Division (CMB) in England stated that they had persuaded British Aerosol Manufacturers Association (BAMA) and the FEA (Federation of European Aerosol Associations) in Western Europe that self-pressurized products placed in their "Bi-Can," a

compartmented can containing an inner plastic bag for the base product, should
not be considered aerosols. This would give them preferred treatment by the
following transportation authorities:

- ADR   European Agreement for the International Carriage of Dangerous
  Goods by Road.

- RID   International Convention Concerning the Carriage of Goods by
  Rail (Berne 1961; Annex 1).

- IATA International Air Transport Association (Restricted Articles
  Board).

- IMCO International Maritime Consultative Organization (United
  Nations).

They asked for industry support in the U.S. No action was taken.

Another definition of an aerosol product that has often been published is
as follows:

"A hermetically sealed metal, glass or plastic container, fitted or
able to be fitted with a valve, and containing a base product and/or
a liquified and/or high-pressure propellant, able to dispense the
contents in a controlled manner as either a spray, foam, stream,
gel, paste, lotion, gas, powder or combination."

In the U.S., for the purpose of interstate transportation, aerosols are
limited to 50 cubic inches (819.35 mL) in metal cans, or to 4 fluid ounces
(118.28 mL) in non-metallic containers. The United Nations recommendation is
1000 mL for all products, and this is generally followed in Europe. In Japan
and other countries, the capacity limit is 1400 mL, although other
restrictions apply. A few countries permit "aerosols" up to 20 liters in
capacity if made of steel. In the U.S., steel cylinders up to 40 liters in
capacity are used for insecticide sprays, egg treating mineral oils, and other

specialized applications. They are not considered aerosols. In some beauty shops, hairspray concentrates are dispensed from pressure tanks maintained at about 100 psig (7.04 bar) compressed air pressure at ambient temperature. The operator uses a thin hose and breakup nozzle for product applications. These products are also not considered to be aerosols.

2. Description of Aerosol Packaging Alternatives

BAG-IN-CAN TYPES

The Sepro Can

In 1954, Croce patented a perfume spray in which the perfume concentrate
and the propellant were contained in separate reservoirs (U.S. Patent
2,689,150). And in 1955, the Metal Box Company, Ltd. was granted a British
patent for a device that would permit the dispensing of flowable products
where the product and propellant were kept separate from one another (Brit.
Patent 740,635). In 1958, the Continental Can Company, Inc. formally introduced their Sepro Can, which contained an accordion-shaped polyethylene
plastic alloy bag inside a specially designed aerosol can measuring 1 1/8" by
6 1/8" (53 x 156 mm). The unit was filled by first adding as much concentrate
as possible to the bag, then sealing the top with a one-inch aerosol valve.
After that, propellant gas was injected through a small hole in the base
section of the can and the hole was plugged with a short length of rubber
cording. Only a few grams of propellant were needed to discharge from 175 to
250 grams of product (according to its density), since the two were kept
separate by the bag. Also, the propellant would never be discharged during
the lifetime of the can, but would remain inside until the empty unit was
crushed, shredded, incinerated or rusted through in a dump site, except for an
infinitesimal amount that might seep through the plugged and double seam can
seals and escape into the atmosphere.

Cross-sectional views of the Sepro Can, a mechanized or pneumatic squeeze
tube, are shown in Figure 2.

285

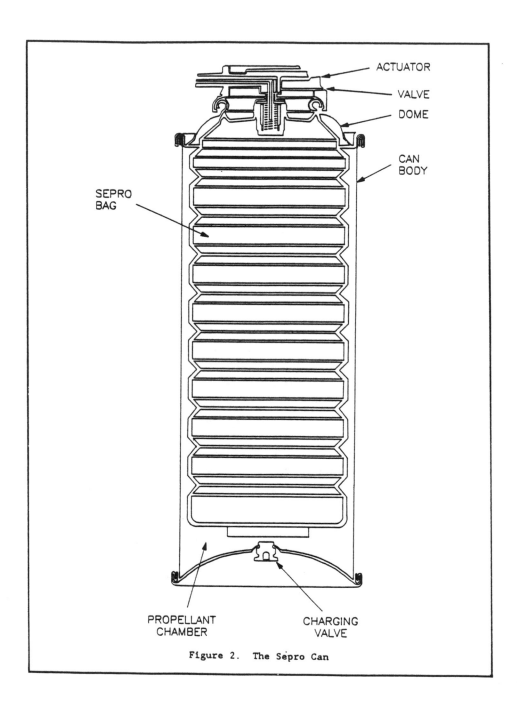

Figure 2.   The Sepro Can

The Sepro Can, from the term "Separate Product and Propellant," was
designed to permit gas-free dispensing and the dispensing of viscous products
that had a positive yield point. It was hoped that this package would
facilitate the growth of the aerosol business by allowing a new range of
products to be packaged under pressure.

The standard aerosol cannot dispense products with viscosities much
beyond 350,000 cps., since such chemicals and formulas usually have a positive
yield point, or, in other words, exhibit shape retention. For example, if a
toothpaste is filled into an ordinary aerosol can and placed under 100 psig
(7.04 bar) of nitrogen pressure, an appropriate valve and spout will dispense
the produce very nicely, although some slight expansion of the extruded paste
may occur as dissolved nitrogen gas slowly forms almost invisible foam bubbles
in the product. (This is too insignificant a feature to be observed by the
casual user.)

Within the aerosol can, however, each actuation causes a further cavitation of the initially flat toothpaste surface. At a certain stage the cross-
section looks as shown in Figure 3. The cavitation area will deepen with each
additional actuation, until it reaches the bottom of the dip tube. At that
point the nitrogen gas will exit in a fraction of a second, and the remaining
product cannot be dispensed.

Toothpastes have been prepared without positive yield points, so that the
cavity left after each actuation will slowly heal--or flatten out. However,
they tend to drip off the toothbrush to some extent and will also leak out of
the valve spout orifice onto the top of the container. Some of the nitrogen
propelled (nitrosol) toothpastes of the early 1960s had spout plugs, connected
to the spout by a fairly thin polyethylene filament. They were designed to be
applied to the spout orifice after actuation, to prevent dripping. Sometimes
the pressure created by releasing nitrogen gas caused them to pop out. One
major marketer (Colgate) kept such a product on the market for about twelve
years, selling 300,000 cans a year to persons who liked the dispensing system.

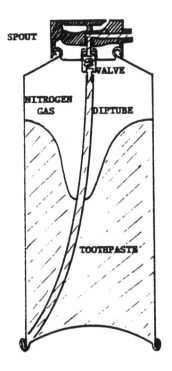

Figure 3.. Ordinary Aerosol Dispenser with Toothpaste.
(About 15 percent Dispensed)

But a slowly diminishing business of this small magnitude was unappealing to Colgate and they finally dropped the item.

In the mid-1960s, the Continental Can Company developed and vigorously promoted to marketers the following four Sepro Can sizes:

202 X 214mm	(3-fluid ounce capacity)
202 X 406mm	(5-fluid ounce capacity)
202 X 509mm	(7-fluid ounce capacity)
211 X 604mm	(16-fluid ounce capacity)*

* Never produced commercially.

A major detraction was the need for marketers to spend about twice as much for Sepro Cans as for ordinary aerosol cans, as well as to install specialized "gasser-plugger" and other equipment on their packaging lines. The package also had a few quality problems. Except for "Edge," a patented gel-type shaving cream developed by S.C. Johnson & Son, Inc. in 1969, no major uses developed.

The evolution of the plugging technology went through several stages. At first, a gasser-plugger would inject a liquified or non-liquified gas into the filled can (1 to 7 g) and then ram the end of a 5/32" (4-mm) diameter lubricated neoprene cord or rod into the 1/8" (3.2-mm) diameter hole in the center of the can base. After insertion, the machine would cut the rubber cord off from the rest of the reel.

These early machines, required to perform three fairly complex operations in a sealed area, were production nightmares and generally the rate-limiting factor in manufacturing operations. Much later, engineering improvements were made to increase the viability of this sealing approach.

Then Continental Can Company announced an improvement known as the Nicholson Model 2 plug valve. It consisted of a solid rubber billet or plug, partly splined on the side wall, which was designed to fit part-way into the can hole during manufacture. In the filling operation, propellant gas was introduced through the splined channels, after which a small ram was used to force the plug fully into the hole, making an hermetic seal. This plug, with only slight residual modifications, is still in use today.

The early plastic bag designs were highly pleated or accordion-walled, and this caused problems when filling viscous concentrates such as pastes and gels. The Continental Can Company purchased a single-head and twin-head Elgin spin-filler, useful for filling these products, which they loaned to certain marketers and larger contract fillers to help in product development work. Larger machines, such as six- and twelve-head Elgins, were available. A very large Consolidated Equipment Company eighteen-head filler was modified for use by S.C. Johnson & Son, Inc. as a spin filler. Finally, the Pfaudler Mfg. Company later introduced a six- and twelve-head spin filler. The Consolidated and Pfaudler machines have a pressurized bowl option, for containing any vapors from shaving gels that contained isopentane $[(CH_3)_2CH-CH_2-CH_3]$ or other flammable or excessively volatile concentrates.

Every concentrate was found to have an optimum spin-filling rate, generally in the range of 400 to 1200 rpm. Below 400 rpm, the centrifugal force was often insufficient to effectively drive the concentrate into the pleated areas, causing unwanted air pockets to form and survive. Over 1200 rpm, concentrate vortexing would exceed gravity and product would be spun upward and out of the container.

During 1971, the products shown in Table 43 were being packaged in the Sepro Can dispenser.

The labeled formulas for the "Edge," "Rise," and "Foamy" gel shave creams are presented in Table 44. In accordance with Food & Drug Administration (FDA) regulations, these cosmetic products must list their ingredients in order of decreasing percentages; those present in concentrations of less than one percent may be placed in any order.

TABLE 43.   PRODUCT MIX FOR SEPRO CAN DISPENSING

Brand Name	Marketer	Product Type	Sales (MM Cans/Yr)
Edge	S.C. Johnson & Son, Inc.	Gel Shaving Cream	11,500,000
Crazy Legs	S.C. Johnson & Son, Inc.	Gel Shaving Cream, for women	500,000
Shimmy Shins	Helene Curtis Industries	Depilatory	100,000
Shimmy Shins	Helene Curtis Industries	Moisturizing Cream	200,000
The Caulker	W.J. Jones & Company	Caulking Compound	Small
Tomato Catsup	Ellis & Associates Inc.	Tomato Puree Catsup	Small
The Meat Eater's Sauce	(Georgia Firm)	Barbecue Catsup	25,000
Steiner's Jewel Gel	Steiner & Company	Ablative Insulating Gel for Ring Repairs	20,000
Natural Honey	(California Firm)	Honey	35,000
Cook's Jolly Jelly	Cook Products Company	Three Jelly Products	30,000
Popcorn Oil	DeLorio & Associates, Inc.	Thickened Corn Oil; in three flavors	25,000
Corn Popper Oil	(Florida Firm)	Soy Bean Oil, several flavors	50,000

TABLE 44.   CTFA LABEL INGREDIENT LISTINGS FOR THE THREE GEL-TYPE SHAVE CREAMS

Edge Ultra Gel	Rise Super Gel	Foamy Shaving Gel
Water	Water	Water
Palmitic Acid	Palmitic Acid	Stearic Acid
Triethanolamine	PEG-150	Oleth-20[a]
Pentane[b]	Diethanolamine	Triethanolamine
Sorbitol	Myristic Acid	Isopentane [b]
Fatty Acid Esters	Isopentane[b]	Lauramide DEA
Isobutane	SD Alcohol 40	Isobutane
Cellulose Polymer	Acetylated Lanolin Alcohol	Peanut Oil[a]
FD&C Yellow No. 10	Isobutane	Mineral Oil[a]
FD&C Blue No. 1	Fragrance	Hydroxyethylcellulose
Fragrance	PEG-90M	Fragrance
	FD&C Blue No.1	Coco-triglyceride[a]
		Menthol
		FD&C Blue No. 1
		D&C Yellow No. 10

[a]Lubricants, used to seal in moisture.

[b]The pentane and isopentane foamants are used in concentrations of about 1.4% of the formula.

The three formulas are quite similar. Each has the di- or triethanol-amine ester of C_{14} to C_{18} fatty acids as the primary surfactant. Sodium and potassium fatty acid soaps are absent, although they are always seen in conventional shave creams. Cellulose, hydroxyethylcellulose, or high-molecular weight polyethylene glycols are used to achieve the gel structure. Most particularly, approximately 1.5% of low-pressure propellants are incorporated into the gel structure, so that the extruded gel can be "magically" converted to a heavy foam on contact with the warm surface of the palm or fingers or if touched and manipulated slightly. The original "propellant" was CFC-113 ($CCl_2F \cdot CClF_2$), but after the FDA ban in 1978 the formula was converted to use isopentane. Later, for a faster and more reliable transition, mixtures of n.pentane/isobutane (80:20) and isopentane/isobutane (90:10) became popular.

Around 1973, a regular shave cream called "Pour Homme" (For Men) was marketed in a Sepro Can. The shave cream contained 3.25% of Propellant A-46 (which is 15 wt % propane and 85 wt % isobutane), and the "exospace" (the volume outside the bag but inside the can) contained Propellant A-60 (32 wt % propane and 68 wt % isobutane). With a pressure ranging from 16 to 29 psi higher than the product at 70°F (1.1 to 2.0 bar at 21°C), the product was extruded at a reasonable flow rate, and since no propellant could ever escape from the concentrate into an expanding head space, as happens with all regular aerosols as they are used up, the foam density and overrun remained exactly constant for "Pour Homme" during its service life. After two years, the company was unable to obtain Sepro Cans to continue its operations and the product was terminated.

At present, the Continental Can Division of U.S. Can Company, Inc. is the only U.S. producer of this type of container. They have a capacity of 50 to 52MM units a year. The cans themselves are produced only on the firm's "Z-bar TFS Conoweld I" line at their fabrication/assembly can-making plant near Racine, WI. The line has a capacity of about 53 million units a year on a ten-shift-per-week basis, and sub-assembly units, such as the pierced base with loosely fitted plug, can be produced at only a slightly greater rate.

The plastic bottle or "bag," as it is generally called, is made at the company's Burlington, WI facility, using three "wheels" or individual production lines, each having a capacity of 17.5 million units a year on a three-shift, five-days-per-week basis. Originally, the bags were made of either polyethylene (LDPE) or of a particular "Conalloy" plastic alloy. The former type has now been eliminated. The current Conalloy bag consists of low-density polyethylene, nylon, and a proprietary binding agent produced by DuPont that makes the two plastics more compatible. The Conalloy bags mold better than LDPE types and have been used continuously since 1968.

While the discontinued LDPE had better resistance to moisture permeation than Conalloy, the latter is superior as a hydrocarbon propellant barrier. This characteristic became critical when Edge shave cream had to be reformulated in 1978 to eliminate the internal CFC-113 foamant and the external CFC-12/114 (60:40) blend ($CCl_2F_2/CClF_2 \cdot CClF_2$) in favor of isopentane and isobutane/propane (87:13), respectively. When hydrocarbon Edge formulas were packed in Sepro Cans with LDPE bags, traces of immediate foaming were seen in gels dispensed after as little as eight months of room temperature storage and five months of 100°F (38°C) storage. In contrast, when Conalloy bags were used, technicians could not detect instant aeration until after 14 to 16 months of storage and the permeability did not escalate to a consumer problem until the product had been stored 20 to 24 months at 70°F (21°C).

During late 1981, a 100% nylon bag became commercially available. It was a stronger, tougher bag than the Conalloy type. It was pre-form or parison molded at a substantially higher temperature than the Conalloy type, making it possible to fill Sepro Cans with very hot products, and even to sterilize them in an autoclave to the usual 252°F (122.2°C) if desired. After cooling to below 122°F (50°C), the Sepro Can may then be gassed and the bottom sealed. Advantages of the nylon bag are that much less plastic is needed, and the critical "T-tab" area at the bottom of the bag knits together very effectively and reliably, like the earlier LDPE bags but unlike the Conalloy type. Voids or thin spots have been a continuing problem of the Conalloy bags.

During 1983, the "Lamicon" bag was developed for Sepro Cans, using a Japanese process for the pre-form or parison blow-molding of multi-layered plastics. The wall of the Lamicon bag is composed of LDPE/adhesive/EVAc/ adhesive/LDPE, wherein the EVAc stands for ethyl vinyl acetate polymer. EVAc provides an excellent barrier for oxygen and other gases. It has a good record in food bottles, and Sepro Can tests show the same good performance. It is also an excellent hydrocarbon gas and liquified gas barrier.

The Conalloy, nylon, and Lamicon bags can be effectively used with a great variety of products, but there are reasonable limitations, as with all packaging systems. Some are listed below:

- There may be problems with products whose viscosities are over 95,000 cps. (e.g., very thick molasses) at the discharge temperature:

  -- Nylon will take up to about 350,000 cps, but transport rate through even very large orifice valves may be slow.

  -- Clayton and Super-Whip valves have "huge" orifices available, if inverted applications are acceptable.

  -- Special Precision, Bestpak, and Beard valves are sometimes useful.

  -- Increasing the propellant pressure outside the bag is useful, up to the practical limit of 70 psig (4.9 bar) at 70°F (21°C). In contrast, regular aerosols can only handle products with viscosities up to 2,000 cps at room temperature.

- Products that exhibit pressure-induced syneresis:

  -- The application of hydraulic or pneumatic pressure to some liquid-in-solid products will cause the liquid to be partially squeezed out of the matrix. An example is peanut butter, where the peanut oil can be synerized out of the mixture, leaving a substantially stiffer bottom layer. Initial experiments have shown that regrinding the peanut fragments will reduce or resolve the problem.

-- When ordinary peanut butters are packaged in Sepro Cans (using nylon bags) the units operate well for the first few days, but then deliver increasing amounts of peanut oil, and after all this is discharged, the remaining paste is too stiff to extrude.

● Products that are highly lubricous:

-- A salad oil product can cause the rubber grommet and seal of the Clayton valve to force out of the valve cup and fly across a room. This does not occur with stem-type valves.

-- A silicone product managed to permeate the bag in micro-gram amounts over many months, ultimately lubricating a nitrile rubber plug to the degree that it popped out of the hole in the base section. Sepro Can storage at 122°F (50°C) exacerbated the problem.

● Products that are acidic:

-- Although vinegar (acetic acid based) products can be placed in Sepro Can bags and used in connection with valve cups that are laminated with polypropylene or lined with nylon, so that no direct metal contact is possible, the contained acetic acid ($CH_3 \cdot CO_2H$) can permeate through Conalloy and nylon bags and attack the tin-free steel (TFS) Conoweld I can surfaces. Reformulation with malic or citric acids is useful in some instances, since these do not permeate to a significant extent.

- Clear gel products that may have long shelf lives before use:

  -- Hydrocarbon propellants are able to permeate Conalloy and nylon
     bags sufficiently to render gel shave cream products sub-
     standard after 20 to 24 months at room temperature. The
     problem seems to be strongly reduced if Lamicon bags are used.

- Certain solvents are capable of bag degradation. A solution that
  included 64% turpentine slowly turned an LDPE bag into a semi-
  viscous mass, and one with a vegetable oil did the same at 122°F
  (50°C). Diethyl ether, an 8% sodium hydroxide oven cleaner base,
  boiled linseed oil, and cyclohexanol all act to slowly degrade
  nylons. The use of alternate bags is sometimes an answer.

- Products that must be hot filled:

  -- None of the available systems will fail when concentrates are
     filled up to about 145°F (63°C) and propellant is introduced at
     any time thereafter.

  - If concentrates are filled between 145 and 160°F (63 and 71°C),
     the propellant must be limited to isobutane, nbutane or their
     mixtures, or over-pressurization will occur. Exception:
     Higher-pressure propellants may be used if the hot concentrates
     are given time to cool to 145°F (63°C) or less.

  -- Above 176°F (80°C) the Conalloy and Lamicon bags may distort or
     have a better chance of dissolving in certain concentrates.
     Nylon bags should be used in such cases, since they can
     withstand temperatures up to 280°F (138°C) with most con-
     centrates.

Depending on bag size, bag composition, and, to a small extent, the
product itself, Sepro Cans will dispense from 94 to 97% of the contents within
the bag. The propellant outside the bag is not dispensed. In the U.S., only

the dispensable amount of the product may be listed as the net weight. As a
rule, the Sepro Can operates with far less propellant than any standard
aerosol product.

Considering hair sprays, most conventional (single compartment) aerosols
use from 22 to 35% hydrocarbon propellant, or about 35% dimethyl ether
propellant in the formulation. For a corresponding Sepro Can of the 202 x 509
size, the amount of exospace propellant would be about 2% of the weight of the
concentrate. Some coarsening of the product would result from this transition, since the usual "micro-explosive" effect of the propellant evaporation
would be absent.

The Sepro Can may be operated in any position, since the bag is always
liquid filled. In fact, during the concentrate filling step, care should be
taken to fill the bag to the maximum, while allowing room for the valve
mounting cup insertion into the throat without overflow. Small air pockets,
when expelled, may sometimes make a "splat" noise and cause some products to
spatter. Fairly costly inverted or "spray-anyway" valve options are not
required for Sepro Cans.

As the product is used up, the bag collapses upward in a controlled way
because of the circumferential pleat design. No bag pinch-off will take
place. The latest pleat profile consists of gently rounded "V"-shaped
indentations between 5/16" (8-mm)-high vertical wall sections. Compared with
the earlier, more sharply indented "V"-shaped pleats, the new bags provide a
more controlled collapse pattern, increased bag capacity and a generally
increased ease of filling viscous concentrates. A minor objection is that the
Sepro Cans become increasingly top-heavy during use, as the bag collapses
upward. This is most noticeable with dense concentrates such as toothpastes.

A special one-inch dome section is required for Sepro Cans, with the
opening enlarged from the usual 1.000 ± 0.004" to 1.021 ± 0.003" (25.40 ± 0.10
mm to 25.93 ± 0.07 mm). This recognizes the approximately 0.10" (0.25-mm)
thickness of the Conalloy and Lamicon bags in that area, so that with bags in
place the net opening will be correct for the standard one-inch valve cups.

In the case of the thinner nylon bags, an intermediate can opening is required.

The Sepro Can may be closed with any standard valve cup and standard crimping tools; e.g., collet and mandrel (plunger). Standard 1.070 ± 0.003" (27.18 ± 0.07-mm) crimping diameters may be used. However, the crimp depth must be made larger, to account for the bag thicknesses at the crown and at the point of hard contact. (Actually, both are crushed to thinner dimensions in the crimping process.) Favored crimp depths are in the 0.180 ± 0.004" (4.57 ± 0.10-mm) range.

The can uses a regular necked-in 201-diameter bottom, uniquely pierced with a 1/8" (3.18-mm) hole in the center and having an upward lip or flange projection into the can. The Nicholson Model 2, two-stage charging valve is supplied by the can-maker and inserted to the first stage (loose) position in the hole. The filler introduces the propellant through the valve's ports, filling the area between bag and can. Depressing the valve to its second position with a ram seals the propellant inside.

The fit between plug and can base is very efficient. The leakage rate is always less than 0.50 g per year, and often less than 0.05 g per year at room temperatures. Should severe over-pressurization take place because of heating, the plug will remain in place even if the can eventually ruptures.

Sepro Cans are equivalent to other aerosol cans in terms of pressure resistance. They can be hot-tanked at temperatures up to 170°F (77°C) if the propellant is isobutane. During hot tanking, the contents of the bag do not significantly warm up, since the gas-filled exospace and plastic barrier are effective insulators. However, because of the pressure exerted on the can by the propellant in the exospace, all DOT regulations are satisfied.

There is a wide choice of propellants in the U.S. The usual ones are isobutane or lower-pressure blends of propane/isobutane. Dimethyl ether might act to soften the bags. For two reasons, the high-pressure compressed gases, carbon dioxide (CO_2) nitrous oxide (N_2O), nitrogen (N_2), and ethane $(CH_3 \cdot CH_3)$ are not appropriate:

> There is no reserve against slow leakage. (In the case of liquified propellants, a small liquid pool is present to replace lost propellant gas by evaporation);

> Pressures will diminish substantially as the bag slowly collapses during product use because the absolute pressure varies inversely with headspace or outage space volume. This can be shown by the following example:

> -- The volume of the exospace is 40 mL. The initial pressure of carbon dioxide is 115 psig. After 80 mL of product has been expelled (about 1/3 of the total), what is the remaining pressure?

>> Initial Pressure is 115 + 15 = 130 psi-absolute

>> PFinal = PInitial $\cdot$ VFinal/VInitial) = 43.3 psi-absolute

>> Final Gauge Pressure = 43.3 - 15.0 = 28.3 psig

> The example shows that such propellants are unable to dispense the entire contents of Sepro Cans.

Practical propellants for Sepro Can gassing are the hydrocarbons: nbutane, isobutane, and propane, or their blends. Establishments unable to safely fill hydrocarbon propellants may wish to use such blends as HCFC-22/142b (40:60) for non-food items. As product viscosity increases, higher-pressure [up to 70 psig at 70°F (4.9 bar at 21°C)] blends of butanes/propane may be preferred. The action of Clayton, Super-Whip, and other stalk-type toggle valves can be stiffer at these higher pressures because of resistance factors.

Sepro Cans are presently being made on only one production line in the U.S. It has a speed of 240 units per minute and averages 110,000 units per shift. It is scheduled to run at a maximum of three shifts one day, two the next, three the third, and so forth, allowing for maintenance during the third shift on every second work day. Considering 200 days of operation a year, the output is 52,000,000 units a year.

Changeovers, from the standard 202 x 509 size to the 202 x 314 sample size, for example, are very costly to the supplier, requiring about 5 days of downtime for both the can-making and bottle-making plants, and a similar period to change back. It would take an order of 5 to 100 million cans for this to be feasible, and the can-maker has so far sold all sample cans to marketers at about twice the usual discounted prices.

Quality problems that caused S.C. Johnson & Son, Inc. to reject as many as 30% of all "Edge" Sepro Can pallet loads during 1980 and 1981 were reduced to a 0.28% reject rate in 1984 and to a 0.21% reject rate in 1987. The quality problems included the following:

- Welding faults--lack of integrity due to the presence of cold weld areas on the can side seam.

- Crimp problems--lack of integrity due to offset parting lines on the bag neck lying against the curl of the can. Since the bag-to-curl interface has to seal pure hydrocarbon gas, with no solvent action to soften or swell the plastic--to help create a more effective barrier--offsetting is a very serious consideration. The offset distance is tightly controlled and measured frequently during bag production.

- Incompletely molded rubber grommet. Grommets are made on a complex, 98-cavity mold, and sometimes the rubber fails to fill the entire volume of each cavity.

- **Weak tail tab on the base of the bags.** Presently a thicker (and thus stronger) knit line is used, together with a recessed construction.

- **Top seam problems.** The top double seam cannot be tested during can production for hermetic integrity, because of the bag. For example, welded end cracking has caused problems in this area.

Despite the manufacturing complexities of the overall Sepro Can package, the Continental Can Division currently feels that it produces a very high-quality item, partly because of the insistence on quality by customers.

The 1989 pricing of Sepro Cans of standard size is as follows:

	100,000 Units	500,000 Units
Base Price	$360.34	$360.34
Coat/Print/Varnish (Outside)	9.89	9.19
Each Additional Print	3.14	2.61
White Dome	2.60	2.60
White Bottom	1.64	1.64

In 1989, there were reportedly twelve marketers or contract fillers in the U.S. who were capable of filling Sepro cans. All but two or three have relatively low-speed Terco, Inc. gasser-plugger equipment, generally rated at 40 cans per minute. One filler (Aerosol Services, Inc., City of Industry, CA) has two such lines.

Since 90 to 95% of the present product mix is the post-foaming, gel-type shave creams, where the concentrate contains 4 to 5 volume percent of very volatile hydrocarbon material [typically isopentane, boiling at 86°F (30°C)], concentrate preparation and filling can be dangerous. The major filler, S.C. Johnson & Son, Inc., chills the concentrate (without hydrocarbons) to around 38°F (4°C) in a closed system, after which the hydrocarbon blend is added. Because of the low temperature, foaming is avoided, even when agitation is applied. The finished concentrate is filled into cans under explosion-proof, well-ventilated conditions. Other methods meter the concentrate and hydrocarbon portions together under cold conditions just before the filling step. The option of adding the gas-free concentrate, and then the pure hydrocarbon blend--or of adding the gas-free concentrate, crimping on the valve, and then pressure loading the pure hydrocarbon blend--is seen as a possible alternative for less gel-structured formulations. These "looser" dispersions would allow intermingling of the hydrocarbon liquid within a reasonable number of days or weeks. Since the bags are completely full, hand or mechanical shaking of finished units is of relatively modest benefit. The process is best served by filling the concentrate (gas-free) in very warm 122°F (50°C) conditions, adding the foamant gas by Through-the-Valve (T-t-V), hot-tanking, and then mechanically shaking cans or cases of cans at the end of the line. The 1989 prices for several Terco, Inc. machines required for these functions are shown in Table 45.

The present sales outlook for Sepro Cans is uncertain. Post-foaming gel-type shave cream marketers are studying the option of a piston version of the same can size, and one has market-tested the concept. Because of the high price of the Sepro Can and the availability of improved versions of the piston can in both tinplate and aluminum containers, most of the limited activity in this field centers in the piston area.

Bi-Can

Around 1987, after extensive research, the Sutton Aerosols Unit, Metalbox Aerosols & Toiletries Packaging Division, MB Group, p.l.c. (England) launched their version of the Sepro Can. Except for a longitudinal bulge, their nylon bag fits snugly to the can body, leaving only a 1/8" (3.18-mm) high space at the top and a somewhat larger volume at the bottom for the exospace propellant. Several can sizes are available. One is 115/114-200 x 515 (50-mm inside diameter by 150-mm inside wall height), while another is 112/113-114 x 312 (45-mm i.d. x 95-mm i.w.h.).

TABLE 45.   PRICES FOR TERCO, INC. GASSER-PLUGGERS (1989)

Production Rate[a]	Type Operation	Cost (Dollars)
40 c/min.	Rotary Indexing	$26,063
40 c/min.	In-line	$45,758
80 c/min.	In-line	$56,355
120 c/min.	In-line	$77,040
120 c/min.[b]	In-line	$88,600

[a]Bottom charging and plugging unit only, except as noted.

[b]Bottom charging and plugging unit, plus through-the-valve gassing of hydrocarbon foamant.

The standard closure is a 3.5-mm pierced hole in the base, which is sealed by a short length of 5.0-mm diameter 80-Durometer nitrile cord after propellant injection.

Before the formal introduction of the Bi-Can at the Interpack trade show in Dusseldorf, West Germany, Metalbox worked with Aerosols International, Ltd. (England's largest contract filler) for over two years getting them ready to produce Bi-Can products on a medium-speed line having a capacity of about ten million units a year. During 1988, the line actually produced about 3.5 million Bi-Can units. As in the U.S., nearly all the business was in the popular post-foaming shave cream gel area, with products by Gillette (well over two million units), Wilkinson Sword, Tesco, Marks & Spencer, and Medicare.

The Bi-Can (short for "Bag-In-Can") is now being promoted for additional applications, including "Nappi" coffee concentrate, two toothpastes, a line of artist's pigments, petrolatums, lithium greases, catsup and mustard, jellies, medicinal liquids, soft- to medium-viscosity caulking compounds, syrups, honey and flavored honeys, medicinal liquids, and cake icings. Baby oils and skin lotions have been demonstrated to customers.

Were it not for the high cost, and sometimes the relatively small size of the package, bag-in-cans might be a very high-volume item. The ability of these units to contain and deliver products is summarized below:

- Can deliver, as well as contain;

- Cannot be spilled;

- Can be made sterile by autoclaving, and will remain sterile during use;

- Can be packed essentially air free for ingredient stability;

- Can handle liquids with viscosities of 1 to 1,000,000 cps. at ambient temperature;

- Can be packed and maintained moisture free (important for moisture-polymerized functional organo-silicones, for example);

- Are not messy to use;

- Are highly directional in application (controlled spray);

- Can be used with container in any convenient position;

- About 98% of the contents can be dispensed--more than many other packs;

- No risk of product contamination by metals, except stainless steel valve spring (The Metalbox "Metpolam" laminated valve cup may be needed);

- Highly concentrated product forms can be dispensed;

- With proper propellant selection, such as nbutane, the package can safely withstand temperatures up to 212°F (100°C);

- Applicable to post-foaming gels, as are the related piston-can and Enviro-Spray bag-in-can packs;

- No concentrate evaporation is possible;

- The propellant is only 1 to 2% of the total contents for most products and is likely to be incinerated with the empty can;

- Can dispense product as a spray, foam, post-foam, liquid, paste, or gel;

- Available in 3- to 9-fluid ounce (30 to 226-mL) bag volumes;

- Bags with as many as four different materials in up to four layers (including aluminum) are available for maximum resistance to permeation, ensuring shelf lives of at least three years in tests to date;

- With the better bag shape and improved filling techniques, thick items can be filled without refrigeration or spin-filling options;

- Low-pressure mixtures of special properties may be packed, such as a water and dimethyl ether mixture, which provides high solvency and soon evaporates completely;

- The package is triply tested: during can-making, when the bag is inserted, and during the bunging or plugging operation;

- A wide range of delivery rates is available, depending on choice of product viscosity, exospace propellant pressure, and valve orifice sizes; and

- A total of 23 "Trimline" sizes are potentially available, from 100 to 1,114 mL.

Metalbox, now actually CMB Packaging, Ltd., formed by the merger of Groupe Carnaud, S.A. and Metalbox Packaging, Ltd. in 1989,, declares that the Bi-Can is not an aerosol. This view is upheld by the British Aerosol Manufacturers Association (BAMA), and several European regulatory bodies (such as COLIPA) that have published conclusions on this subject. Bi-Cans are presently being shipped within the United Kingdom as non-aerosol commodities.

Compack

Around 1973, Aerosol Services Moehm, S.A. of Switzerland (now ASM, S.A.) developed a Lechner™ aluminum aerosol can with a LDPE vertically pleated bag and a valve cup lined heavily with plastic. Several sizes were offered, with diameters of from 1 3/8" to 2 1/8" (35 to 52 mm). Product leakage at the

308 Alternative Formulations and Packaging to Reduce Use of CFCs

complex interface of can bead, bag flange, and valve mounting cup was a problem. Products such as Blendex toothpaste and a paste-type shampoo concentrate lost marketshare and this forced the company to look for refinements.

Alucompack

This bi-compartmented aerosol used a very thin aluminum tube with a flat base and flanged top as the inner container. This eliminated the gas permeability that had plagued the previous design, the LDPE bag. Also, if the system was combined with a seal of epoxy resin at the crimp area, using air as the propellant, it offered a three-year shelf life, guaranteed by the supplier.

The inner tube, or "Alu-Bag," of D-1 (Heat-Killed, minimum temper) aluminum was generally 0.992 ± 0.003" (25.2 ± 0.08 mm) in diameter and typically 6" (151 mm) long, or about 1/2" (12.7 mm) shorter than the outer aluminum can. Under the 3/32" (2.4 mm) top flange there is a thin neoprene rubber gasket, which is the area that can be improved by sealing with epoxy. The valve cup is fitted with 0.040" (1.00 mm) neoprene or buna-N rubber gasket, called a cut (or lathe cut) cup gasket, since this affords a more reliable seal than the Weiderholder or other Flowed-In, water-based neoprene gasket types.

The aluminum inner tube offers more resistance to pneumatic crushing than its plastic counterparts; therefore, it is necessary to use a fairly high-pressure propellant in the exo-space between tube and can. At least 28 psig (2.0 bar) of pressure differential should be available or dispensing will be incomplete. The aluminum tube is (rarely) susceptible to a kinking type compressive distortion; therefore, it is useful to insert a length of polypropylene capillary tubing in it before filling with the concentrate. Typical dimensions are 90 to 95% of the length of the tube and 0.090" o.d. by 0.060" i.d. (2.29 x 1.52 mm).

The i.d. of the outer can is typically 30 to 35 mm, which means that the volume of the exospace is greater than that of the inner tube or bag. Air or nitrogen pressure can then be used, and the pressure will not decrease substantially during tube collapse, unlike the situation with Sepro-Cans and Bi-Cans. Since less than 1.0 g of air or nitrogen is used, the degree of hermetic sealing must be very good.

The Alucompack development was used for toothpastes, a caulking compound for bathroom crack and crevice filling, and three medicinal items. Aerosol Services, A.G. performed the filling. They developed a technique of adding hydrocarbon propellants, first strongly cooled, into the outer can.

Of the 3 to 4 grams poured into the can, perhaps 1.0 to 1.5 grams evaporated before the evaporation stopped. Meanwhile, the inner tube, smeared with epoxy, was slid through the one-inch (25.4-mm) opening and quickly filled nearly full with product. The capillary tube was inserted. The valve, without dip tube, was placed in position and hermetically sealed. When pouring isobutane, propane, and their mixtures into cans, these dispensers were never filled outside of Europe. Approximately 1.7 grams of liquid hydrocarbon are released when Under the Cup type gassers release aerosol cans. and this is the major charging method employed in the U.S. and Canada.

Micro-Compack

The third variant, by ASM of Switzerland, is a smaller version of the Alu-Compack, generally holding about 10 to 15 mL of product. Such products as a "small area" depilatory (for facial hair, moles with hair, etc.), an anti-wrinkle (Retin A) cream, and various medicinal ointments are filled in these small dispensers. The 13- to 20-mm diameter ferrule-type valve is attached by standard 18- to 24-tine mandrel clinching techniques.

Lechner (Types I Through IV)

The Lechner GmbH System I can was developed around 1974 and consisted of a vertically pleated LDPE bag in an aluminum can, sold with or without a nitrile rubber plug in a 3.5-mm hole in the base. The filler can pressurize and seal the can with a gasser-plugger machine, or, with the plug in place, with a syringe needle that penetrates the plug. The Aerofill, Ltd. firm in England is one supplier of syringe gassers; they claim their hardened steel needles can last for filling up to 60,000 cans. The needles are eventually weakened by dulling and abrasion from the filler substances in the nitrile rubber plugs.

System I is limited to LPDE, and (now) HDPE (D1018) bags in seven sizes ranging from 50 to 400 mL. Such products as antiseptic sprays, household and car cleaners, medicinal and veterinary products, contact lens cleaners, disinfectants, depilatories, and air fresheners have been sold in these packages. Despite the sales efforts of Lechner USA Ltd., nearly all these products are sold only in Europe.

The System II dispenser has been more successful. It uses an inner aluminum bag and outer aluminum can, with top double seam and a dome that may be either aluminum or tinplate. The extruded, cut off and flanged outer can and inner bag are fitted together and triple seamed to the dome. It uses base gassing as described for the System I unit. Introduced around 1978, it is available in 14 sizes (18 bar), two at 15 bar (217.5 psig minimum burst), and one at 12 bar (174 psig). The last size is the largest: 502 mL, which was originally conceived for holding highly acidic hair coloring paste and for use with very high-viscosity silicone-based caulking compounds for commercial uses, but it is now used to dispense a wide variety of products. One interesting use for both this and the Alu-Compack is as the power unit for a nail dispenser. A particular gas blend of allene and methylallene, having a pressure of 90 psig at 70°F (6.33 bar at 21°C), is filled to a capacity of 100% into aluminum bags surrounded with propane, which has a pressure of 108.5 psig at 70°F air free (7.64 bar at 21°C). A micro-metering valve located outside the can feeds a tiny amount of "MAPP Gas" to the firing chamber of a

very small spring-loaded ram cylinder when a triggering mechanism is pressed.
A minuscule spark plug explodes the mixture above the piston-ram, and the ram
then moves outward to drive in a nail, separated from a nail-pack and held in
position below the ram.   In this way, nails can be driven into wood as fast as
one can pull the trigger.. A 3.99-fluid ounce (117.98-mL) charge of MAPP Gas
can sink thousands of nails.

A number of topically applied medicines and drugs are under test in
small versions of the System II dispenser. The 3M, Inc. firm uses it for
their fuel injector engine cleaner, since it is important that only the liquid
phase enter the engine for spark plug and upper cylinder area cleaning
purposes. Prescription dental gels and gum cements are conveniently dispensed. One new drug under development in Europe requires autoclaving at
275.4°F (135°C). After the filled System II unit is processed, heated, and
cooled, it is gassed and plugged.

The Lechner System III dispenser consists of a conventional aluminum can,
but heavily lined in such a way that the lining adheres tightly only at the
crimp and upper dome area. The rest is very loosely attached. When the can
is filled with concentrate and sealed, gas is injected through the hole in the
bottom, causing the bulk of the lining to separate from the can surface and
become, in effect, a bag. The base hole is then plugged. (Syringe filling
cannot be done for fear of perforating the adjacent bag material.)

The modified polyolefin lining is suitable for a wide range of products.
In fact, virtually any cream, gel, lotion, ointment, paste, or liquid now
packed in a plastic tube or bottle can be more conveniently packed in the
System III. At least 98% will be discharged.

Finally, the Lechner System IV is a modification of the System II, that
improves on the relatively poor aesthetics of the triple-seamed dome design.
It looks like a standard aluminum (one-piece) aerosol can. The larger version
with a 1" (25.4-mm) opening will be available by October 1989, and smaller
ones, using a 20-mm ferrule type valve, will come onto the market about March

1990. They will have a 0.98" (25-mm) diameter at first, but other diameters will be available later on.

The most significant of the Lechner developments is the System III dispenser, since it eliminates the preformed bag as such, and thus eliminates about a third of the component cost. This considerable economy should do much to stimulate volume growth of the two-compartment dispensing system.

Presspack

In 1976, this vertically pleated medium density polyethylene (MDPE) bag in aluminum or tinplate can development was completed by a West German firm in Hamburg. By the following year, two West German and one French firm announced they were ready to supply it commercially. However, because of seepage problems around the crimped seal (especially in the case of the aluminum cans), the dispensers were sold in relatively small quantities for about two years and then discontinued.

Other Bag-In-Cans

A number of additional bag-in-can designs have not achieved commercial success, sometimes in spite of excellent designs. One of these, developed in California and taken over by the (then) American Can Company, used a "cup" of polyethylene or laminate structure, attached via the top seam of a regular tinplate can. Production problems consisted of trying to uniformly air blow the plastic cups into waiting "domeless" cans, and trying to eliminate the "Z"s or "switchbacks" that occurred over the flange. When these triple thickness areas of plastic were wrapped into the top triple seam, actually a sextuple seam resulted, which leaked slowly or latently at the fold inter-faces. The companies eventually halted product development.

PISTON CANS

As with the Sepro can, this compartmented package requires less propellant than a conventional aerosol package. Probably the first commercial
piston can was developed in 1961; it contained about 4 Av.oz. (113.4 g) of
Brylcream in an aluminum two-piece container with a free piston of medium-
density polyethylene (MDPE).

Until about 1988, there was only one piston supplier in the U.S., the
American Can Company (now the American National Can Division of Pechiney,
S.A.), who supplied a two-piece aluminum Mira-Flo container. There are
several in Europe and at least one in Japan.

During 1974, a drawn-and-ironed two-piece steel can was manufactured for
the U.S. Borax and Chemicals Company's "Boraxo" waterless hand cream. The can
was a seamless steel type, with top double seam. The problem with soldered
and other welded side seam steel cans was that the polyethylene pistons could
not fit the can wall snugly in this side-seamed area, and this caused "blow-
by" of the gas past the piston wall and into the product, where it usually
dissolved and lost its pressure. In fact, aluminum cans with wall dents were
probable candidates for blow-by problems. During 1986, innovations in side-
seal technology created the smooth side-seam profile, improving on earlier
constructions of the "stepped" type.

The tinplate or tin-free steel (TFS) cans are probable candidates for
piston can modification. They are approximately 20% less costly than aluminum
and are available in sizes of 100-mL to 1114-mL total capacity. The metal is
also harder and less vulnerable to pre-filling or post-filling denting
problems.

The Mira-Flo Can

Experimental piston cans date back to 1956, when Crown Cork & Seal
Company used a crude piston, over a large, compressed, steel spring to
dispense various food products from their 202 x 406 Spra-tainer can. The

spring provided backup pressure in case of propellant leakage. American Can Company began work on a two-piece 202-diameter aluminum can about 1960, and in 1962 they introduced their Mira-Spray (single compartment) and Mira-Flo (piston-type) 202 X 406 containers. The Mira-Flo had a 9/64" (3.57 mm) punched hole in the base, suitable for gas injection followed by plugging with a cord of 70-Durometer neoprene rubber.

The new cans were made at a plant that had an initial capacity of 38,000,000 units a year. Since the price of the Mira-Spray can was somewhat higher than that of the highly similar, steel, two-piece Spra-tainer made by Crown, sales were very poor.

American Can Company promoted the Mira-Flo cans, which were made on the same production line as the Mira-Spray cans, except that an LDPE piston and pierced base section were inserted. Samples containing domestic and imported cheeses were shown to Kraft, Nabisco, General Foods, and others from 1961 through 1963. Samples containing oleomargarine and thick chocolate syrup were shown to Mazola and Bosco product managers at the Best Foods Division of CPC International Inc. An experimental margarine sample, prepared at the American Can Company's Barrington, IL Research Center for the Land O Lakes Co., Inc. survives today, works well, and still contains product that has a good, fresh taste after 27 years.

After a few years, the capacity of the production line was reached: approximately 1,000,000 Mira-Spray cans for various small uses (such as air fresheners), 33,000,000 Mira-Flo cans with various types of cheese, and 4,000,000 Mira-Flo cans with various colors of Pillsbury's Cake Topping for creating decorative designs and/or messages on cake icings. The toggle-action Clayton Corporation food valves were supplied with several alternative actuator tips (in the case of the cake toppings) to create extrusions with star shapes, ovals, etc., in addition to the standard round ribbon. Instead of the more costly aluminum option, American has turned to the welded tinplate piston can.

While many fairly complex piston profiles have been developed (mainly in Europe) over a thirty-year period, in 1987 American pioneered the "free-floating type: a piston that had to expand slightly to reach the can walls. It requires the product to have a viscosity of over 12,000 cps at ambient temperatures. The novel pistons use a "gasket" formed by the product itself to effect the seal and inhibit gas by-pass.

The second innovation was commercialized in late 1986. Known as the "umbrella valve," it is a form of rubber plug shaped like a mushroom. It is applied from inside the can bottom. Small ears prevent the plug from being pushed into the can. The larger umbrella top is flexible enough to compress upon insertion. Once the plug is in the can, the internal pressure forces it downward, making a tight seal. The new valve allows filling speeds of up to 360 cans per minute.

Disadvantages of piston cans are that they require products of reasonably high viscosity that do not distort the piston. Piston by-pass (or blow-by) and permeation can cause problems by reducing the quantity of gas below the piston and perhaps by causing foam generation in the product. If the product is not compatible with the can, this can be a bigger problem with piston cans than with bag-in-can types because of the direct exposure of can surfaces to the product. New, very heavy linings are being developed by CMB Products, Ltd. and others to counteract this shortcoming. Some of the linings are bonded polypropylene approximately 0.010" (0.25 mm) thick, and they can also be used for the double seam sealing material.

Other Piston Cans

Piston cans using aerosol containers have been marketed by Advanced Monobloc, Ltd. (Division of CCL Industries, Ltd., Toronto, Canada), the Continental Can Group of United State Company, Inc., Boxal/Alusuisse, Cebal/Pechinery, S.A., Hoell, GmbH (Hamburg, West German), Rocep Pressure Packs, Ltd. (Glasgow, Scotland) and a firm in Japan. Except for the Rocep units, most are typical piston cans and conform to the description of "Mira-

Flo" units. Perhaps the largest manufacturer is the U.S. Can Co., since they now have a considerable share of the shave cream business.

Rocep cans are unique in that they are often quite small, such as 1" (25 mm) in diameter, and sealed with a 22-mm type ferrule valve. However, the main difference from others is that they use a double piston: two rather shallow types, one below the other, with the small space in between filled with mineral oil. The purpose of the oil is to capture and retain any propellant that penetrates through or around the lower piston, so that it will not go further upward and get into the product. Another unique feature is the "lever pack" package design, where the valve turret head is turned varying degrees on an eccentric track to control dispensing rates. Dispensing can then be actuated at various rates by depressing a 3 to 4" (76 to 101 mm) wire profiled lever against the can. This type of control is perfect for sealants, among other products.

Silicone acetate or silicone aminofunctional caulking compounds, which turn to a rubbery mass when brought into contact with humidity or moisture illustrate this product type. These products cannot be allowed to contact the liquified propellant or various degrees of foaming would take place as they were dispensed, leading to a strange-looking and less-effective seal. Under such trade names as "One Tough" Silicone Sealant blister packs containing three-ounce (80-g) piston cans are being sold at $5.95 to $6.95 each. Standard trigger-type caulking gun packs with twelve times as much of the same product are sold for $4.00 to $4.50 in the same stores.

An interesting alternative to the piston can is a pressurized pack designed to clamp onto one end of a standard caulking canister, filled with organo-silicone, acrylic, butyl or thiokol sealing compounds. An independent plastic piston is pressed into the cylinder by finger-pressure on an actuator. When the finger pressure is released, the gas pressure is automatically discharged upward, out of the orifice in the actuator. This instantly stops the flow.

The Rocep package uses a nonflammable mixture of HCFC-22/142b (40:60) for a pressure of about 83 psig at 70°F (5.85 bar at 21°C), including the partial pressure of trapped air.

A major problem developed when the relatively high pressure and solvent action of the propellant blend softened and squeezed the neoprene plugs out of the pierced holes in the tinplate can bottom, thus depressurizing the units. Short of using CFC propellants, there is no lower-pressure, nonflammable propellant currently available. The issue was finally resolved by moving to a slightly thicker nitrile plug and punching the hole in such a fashion that a slightly ragged lip was formed at the inner rim of the hole to act as a barb. Even so, extended storage at 130°F (54.4°C) will cause expulsion of the plug. Recent findings suggest that a trace film of mineral oil (from between the pistons) is a contributing factor.

The final difficulty with this package is that the container itself is composed of a special aluminum top and wall extrusion, but the base is of tinplate. An uneasy junction of these two metals at the bottom double seam is produced after the silicone product and double piston assembly are added. The tinplate tends to cut into the much softer and thicker aluminum, resulting in a 6.2% leakage rejection rate at the factory hot-tank tester. Technology exists to double-seam dissimilar substances, even plastic container walls to tinplate end sections, but it is not very effective for small-diameter closures such as the 1" (25.4-mm) diameter container.

Apart from the silicone-based specialty bathroom tub and tile sealants just discussed, no other products have yet been commercially produced in this packaging form. The high production cost, partly due to high scrap rates, is considered to be a major factor.

The Boxal Pump Dispenser

During the International Packaging Exhibition (Pakex 89; Birmingham, England; April 21, 1989) the Boxal Group, a member of Alusuisse Packaging

Division, showed their standard piston-type aerosol can, as well as a new, propellantless version that operates on a vacuum suction principle.

Using a custom-made valve by Coster, S.A. and a standard aerosol-type aluminum can with inner piston and perforated bottom (but no plug), the unit provides a metered flow of product whenever the actuating spout is operated. Lotions, creams, and pastes may also be dispensed.

When the pump actuator is depressed, a low vacuum is created in the product compartment. Through the hole (1/16" or 1.6 mm) in the base, atmospheric pressure then presses the HDPE piston upwards a small distance. The pump is constructed to prevent any contact between product and air until the material is discharged from the dispenser.

The pump stroke volume can be adjusted according to product characteristics and according to marketer requirements. Partial strokes will extrude correspondingly less product than full strokes. Because the pressure differential (between a partial vacuum and normal air pressure) is relatively low, the unit is not suitable for highly viscous products. They would emerge too slowly for customer satisfaction.

The development is not inexpensive, due mainly to the cost of the specially designed Coster, S.A. pump-action valve. It does provide a new and attractive packaging form for creams, gels, pastes, lotions, and other low-to-medium viscosity products in the cosmetics, toiletries, pharmaceutical, and food areas, usually giving those products prolonged shelf lives in comparison with packaging in jars or bottles. Evaporation, air oxidation, fragrance deterioration, spillage, and breakage are all avoided. The products can be filled on standard aerosol equipment at high speeds. From 96 to 97.5% of the material can be dispensed. In the U.S., unlike aerosols, the actual content (in fluid ounces), rather than the dispensed weight, is the basis for the declaration of contents on the label. However, labeling requirements will vary with the country having jurisdiction.

In one instance, an aqua-colored, rather viscous specialty shampoo was found to change color rapidly, toward green, then olive green, and finally yellow when packed in glass and exposed to sunlight. No color change has yet been seen in this product when packed for eight months in the Boxal pump dispenser.

INDEPENDENT BAG-IN-CAN SYSTEMS

During the late 1950s, inventor Ellis Reyner began to introduce aerosol marketers and fillers to his patented process for a product designed to permanently separate the propellant and product. In its simplest form, his innovation consisted of a plastic pouch, to be inserted in an aerosol can, either before or after filling with the concentrate. The pouch contained two chemicals: sodium bicarbonate ($NaHCO_3$) powder and a 50% solution of citric acid [$C_3H_4O(CO_2H)_3$] in separate burstable tubes. When the two chemicals come together, during a deliberate rupturing process directly ahead of the valve insertion and sealing operation, they chemically react to produce various sodium citrate salts and carbon dioxide (CO_2) gas. The outer envelope of the pouch remains intact, but it swells as a result of carbon dioxide pressure and presses against the can and the contents, so that when the valve is actuated, product flows out of the can as a coarse (non-aerated) spray, as a gel, paste, post-foaming gel, stream, or foam.

A problem with the early developments was that the bag was subject to gas permeation, stress cracking, product influences, and imperfect welding. These problems were solved by using laminates, often including a core layer of 0.0005" (0.013 mm) aluminum foil to almost totally eliminate any permeation. A less-effective barrier material is Mylar (polyethylene terephthalate - biaxial), which also adds considerable strength to the bag.

A second problem was that the bag could initially swell up only to the volume of the gas space over the concentrate. Because of various government regulations limiting aerosol pressures to about 180 psig at 130°F (12.68 bars at 54.5°C), the practical maximum pressure that the bag could exert at room temperatures was 142 psig (10.0 bars), and many marketers were more comfortable with 60 to 80% of this level. Following is an example of the pressure decrease during use that would take place for a typical product:

Product: Toothpaste.

Volume Fill: 250 mL of toothpaste, 10 mL for pouch, and 140 mL for head space over the toothpaste.

Pressure: 140 psig at ambient temperature - initial. (9.86 bars)

Note: Head space air compression to about 10% of original volume, plus absorption of some of that into the tooth-paste, is not considered here. (Can be reduced by vacuum crimping.)

After the essentially complete discharge of toothpaste, the pouch volume will increase from 10 mL to 400 mL.

After the initial step of gas formation in the bag, it swells to 140 mL and has a pressure of 155 psi-absolute at ambient temperature.

Using Boyle's Law, the pressure drop during toothpaste expulsion will be:

$$P_f = P_i \ (V_i/V_f) = 155 \ (140/400) = 54.25 \ \text{psi-absolute}$$
$$P_f = 39.25 \ \text{psig} \ (2.76 \ \text{bars})$$

Thus, the gauge pressure at ambient temperature is reduced from 140 to 39 psig (9.86 to 2.76 bars).

Repeating this study using an initial gauge pressure of 100 psig, would result in a final (can empty) pressure of only 25.25 psig (1.78 bars). This degree of pressure drop will result in significant decreases in delivery rate, especially for viscous products of positive yield point, during package life. This drop can be reduced by using what has been termed a "functional slack fill" of product, such as a 50-volume percent quantity, but this increases the cost per unit weight or volume of product and has other disadvantages.

Around 1975, the Grow Group, Inc. became interested in the Reyner system, thinking it could be refined to deliver a certain amount of gas at the onset, and that maintenance amounts could be provided as needed during use. After a research period lasting two years, the Grow Group announced the acquisition of Reyner's interests and the formation of Enviro-Spray Systems, Inc. to promote and sell the improved pouches and filling technology that had been developed.

The pouch now contained one fairly large inner container of 50% citric acid solution in water, plus six other much smaller containers. The larger receptacle could be torn or ruptured, either by striking the inserted bag with a small ram, or by the action of the vacuum crimping operation, releasing the contents and generating from 60 to 100 psig (4.2 to 7.0 bars) of carbon dioxide gas. As the product was used, the bag expanded as the head space expanded, and at a pre-engineered point the first smaller compartment of citric acid solution was breached. This returned the pressure to the original level. The process was repeated until the last citric acid receptacle had been ruptured.

With this sort of arrangement, the pressure could go (for example) from 100 psig to 80 psig, back to 100 psig, down to 77 psig, up to 103 psig, etc. as many times as there were citric acid receptacles. The relative complexity of having pressures of over 60 to 80 psig (4.2 to 5.6 bars) was questioned during the development of this system, as was the need for six maintenance system bags. Four of these appeared to be adequate, and future editions of the pouch ultimately used only four.

Other refinements include adding a flow tube, which consisted of a suitable length of aerosol valve dip-tubing so that the expanding pouch would not press hard against the middle or upper potions of the can wall and cut off or trap product below that point, keeping it from being discharged. Finally, the reservoir of sodium bicarbonate was contained in a water-soluble polyvinyl alcohol plastic, so that when the water-impermeable membrane between the primary citric acid sack and sodium bicarbonate compartment was deliberately ruptured by ram or vacuum action, the pouch would not instantly inflate, but would be delayed for one minute to allow time for the valve crimping (or sealing) operation.

Since the bag would be expected to inflate after the package was sealed, a way had to be found to authenticate that it had actually expanded. On production lines this was done by means of X-ray based level measuring equipment. Pressure measuring could be performed on a laboratory or statistical production quality assurance scale, and this also showed if only the main citric acid receptacle had ruptured. One problem with the system, even in units produced in 1989, is that two or more of the citric acid containers can rupture if there is a problem with bag quality, resulting in excessive internal pressures.

Figure 4 depicts a slow-speed aerosol line, using semi-automatic pouch-stuffers, rated at about 18 units per minute for each of the two machines in use. The rest of the line is fairly standard, except for the level checker, which is used to ensure pouch inflation.

The preferred valve is the Precision Valve Corporation Model 1-NN, with a 2 X 0.5-mm stem slotted "Enviro-Spray" type housing. Any type of actuator button or spout may be used. The standard pouch is designed to be used with a 202 X 514 (53-mm diameter X 300-mL) can with a 170-mL product fill.

The firm suggests the following product possibilities:

- Air Fresheners;
- Plant Sprays:
  -- Leaf Shines,
  -- Aphid control,
  -- Fertilizer Concentrates;
- Petroleum Jelly (for example, for babies);
- Bathroom Cleaners;
- Toothpaste;
- Post-foaming Gels (as shave creams);
- Metered Dispensing (micro and macro);
- Toppings;

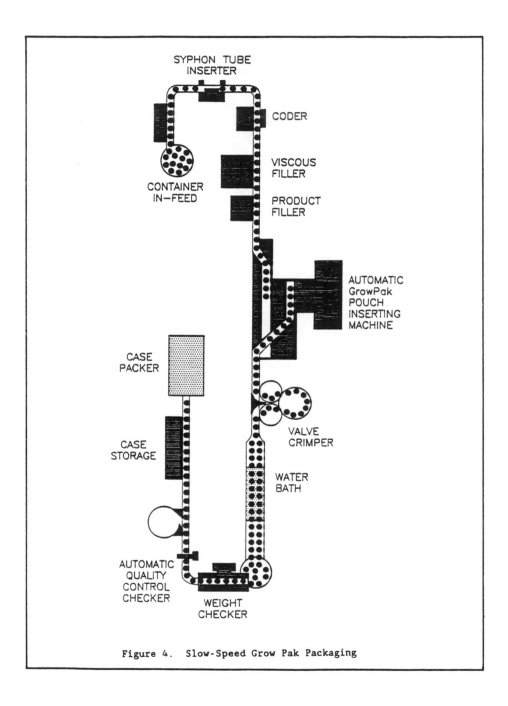

Figure 4.   Slow-Speed Grow Pak Packaging

- Cheeses or Snack Items;
- Waterless Hand Cleaners and Related Lotions;
- Pet Care Items;
  - -- Groomers,
  - -- Shampoos - optionally insecticidal,
  - -- Flea & Tick Sprays (soundless);
- Cake Decorations;
- Industrial Maintenance Items, including lubricants;
- Selected Coatings;
- Furniture Polish (in lotion forms); and
- Mustard, Catsup, Purees and so forth.

Those products actually marketed in the Enviro-Spray System include the following:

- Tomato and Vegetable Insecticide;
- House Plant Insecticide;
- Rose & Flower Insecticide;
- Flea & Tick Spray for Dogs;
- Flea & Tick Repellant Spray for Dogs;
- Spray for Cats - Insecticide;
- Leaf Shine for Ornamentals;
- "Le Gel" by Williams (Beecham) Shave Cream;
- "Kouros" by Yves Saint Laurent;
- "Algipan" by Labaz Sanofi- Rubifacient Cream;
- "CCRF" Tomato Paste, Tomato Ketchup, and Mustard;
- "Mist & Feed" Foilant Nutrient Spray; and
- Beecham Caovel Pet Insecticide Spray.

In 1986, costs were $3.59 to $8.99 for cans ranging from 7-Av.oz. (200-g) to 32-Av.oz. (946-g) net weight. Containers were also sold for such specialties as institutional "gallon-size" insecticides, the pressurization of low-gas beer kegs, soft drink dispensers designed to operate under "no gravity" conditions in the NASA space program, etc. The pet sprays benefitted from the soundless delivery of the Enviro-Spray dispensers, since pets can hear and are distressed by the very high-pitched sound of standard aerosol sprays, except for those pressurized by air or nitrogen gases.

The second largest Enviro-Spray in Europe is a line of four food products by C.C.R.F. (France), under the brandname of Claude Vetillard. They include Tomato Puree, Double Concentrated 28% Mustard "Forte de Dijon," and Tomato Ketchup, packed to 280 g (250 mL) in metal box "Slimline" cans measuring 57 X 164 mm. Each is fitted with a Precision valve and "captured plug" spout. After 27 months, some cans of the Tomato Ketchup have shown a slight seepage of the product at the juncture or top and side seams. With appropriate adhesive-backed formula and precautionary stickers, these cans have made a modest entry into the more expensive U.S. specialty shops, such as those at airports and major hotels.

PUMP SPRAYS - ASPIRATOR TYPES

Pump-sprays have taken many forms. There are those whose pressure is generated within the meter-spray valve, and others (much rarer) whose pressure is produced in the container by various means. In the unique "Pre-Val" unit, developed by the Precision Valve Corporation (1975), a glass or plastic jar is filled with product and then sprayed out by aspirating it up a dip tube leading into an upper "Pressure Pack" containing a liquid propellant. When the valve button is depressed, propellant gas is discharged, sucking up a certain amount of the concentrate and discharging it as well. The usual ratio is about 4 to 1, so that approximately 400 g of concentrate is dispensed by 100 g of a hydrocarbon blend. No solubility of propellant and concentrate is necessary. If the concentrate might dry in the valve orifice to form a solid clog, or after use, the jar portion can be disconnected and the valve actuated to blow the mechanism essentially free of all product. The dispenser, along with refill units, can be purchased in hardware stores, lumber yards, and similar outlets.

The original form of the aspirator-type dispenser is the pump-sprayer for space spray insecticides. This normally consists of a tubular barrel (the cylinder) and a thin piston at the end of a fairly long rod, as shown in Figure 5.

Figure 5.   The "Flit Gun" Aspirator-Type Insecticide Space Sprayer

The product is aspirated up a plastic or metal dip tube and through a jet orifice that ends in the midst of a vigorous stream of air, at 10 to 20 psig (0.70 to 1.41 bar), directed at it from a nozzle at the end of the cylinder. The ratio of low-pressure air to amount of aspirated product is the primary determinant of particle size distribution.   Ideally, the particle size would be less than 30μ. Otherwise, the larger particles would fall to the floor rather quickly and be of little use in killing houseflies, mosquitoes, and other flying insects. The largest-selling insect sprayers have been a line called "Quick Henry, the FLIT," sold by Penola Oil & Chemical Corporation, and later by Esso Oil Company, Humble Oil & Refining Co., and still more recently by Exxon, Inc.

These sprayers were often manufactured in very expensive forms, such as in nickel-plated bronze, with decorative designs and printing (sometimes engraved), and with small boxes of replacement piston "leathers" and spare glass jars that were often customized and suitably embossed with the name of the sprayer. Quart (946 mL) cans of insecticides in low-odor kerosene solvents were available from Penola, Esso, Sinclair, Phillips, Conoco, Shell, Peneco, Gulf, Pennsoil, Rex, Sohio, Cook, and other oil companies, which would work well in any of the available sprayers.

Today, a few firms make all-plastic sprayers, except for the metal orifice areas, but they are not advertised, and sales volume is limited. "F"-style or cone-top cans of insecticidal concentrates can also be found, but availability is also limited. These sprayers are far more popular in countries other than the U.S., Europe, and Japan.

The aspirator-type sprayers are the only sprayers, other than aerosols, that can produce a space spray. They have been so closely associated, however, with (smelly) insecticides that it would not be possible to market them for air fresheners or for other uses in the U.S. However, some "mini"-sprayers of this type are occasionally available for household perfumes in Latin America, and the rubber-bulb type aspirator may still be seen for personal fragrancing applications, generally in rather fancy designs. Colognes are available in bottles exceeding one U.S. gallon (3,786 mL) capacity for refilling other containers and dispensers, so a supply of the product itself is not a problem.

PUMP-SPRAYS - STANDARD TYPES

The Finger-Pump Sprayer

The most common pump sprayer is commonly called the finger-pump, mainly to distinguish it from the trigger-action sprayer. The finger pump is available from the same manufacturers that produce aerosol valves, such as the Calmar Corporation, Bakan Products Co., Risdon Manufacturing Co. (Division of CMP Products, Ltd., as of 1989), Emson Research Company, and others. The largest is probably the Seaquist Closures Division of Pittway, Inc., located in Cary, IL (U.S.). In many cases, it takes an expert to distinguish between a ferrule-type aerosol valve and a ferrule-type finger-pump valve; they are often made by the same company and have two or three components in common. Distinguishing features of the finger-action valve are its larger, more complex valve structure and its often clear plastic body component that displays a complicated spring above a metallic ball check unit. The best indicator is the type of container. If it is a plain glass bottle larger than one fluid ounce (29.57 mL), or a small glass bottle with flat surfaces or sharp corners, or a polyethylene or polypropylene or vinyl bottle, or if the bottle can be deformed by squeezing, or if the valve is attached by means of a screw-threaded connection, the valve is not an aerosol valve. Some finger-action valves are placed in one-inch aerosol cups and crimped onto aerosol cans. These defy identification except by operating them.

The usual aerosol valve has up to seven components and sells for about $40.00 per thousand. In contrast, the finger-action valve has eleven components, some of which must fit together with tolerances more demanding than those of aerosol counterparts. Consequently, these valves sell for two to three times the cost of the aerosol types, depending on size and other factors.

An illustrative sketch of a typical screw-cap mounted finger-action valve is shown in Figure 6.

The finger-action valve delivers a fixed amount of product per actuation, from 125 to 200 microliters. To convert this to milligrams per shot, simply multiply the microliter rating by the product density. Densities may vary by 20% or more. Ethanol solutions, such as hair sprays, have the lowest density, at about 0.80 g/mL.

As the actuator is depressed, the adapter and stem components are forced downward as well. The stem travels a fixed distance into the body chamber, which is normally filled with product in the primed valve. The product forces the piston to expand outward, allowing product to flow past it and into the cross-hole orifices of the stem. From there it travels up the stem hole, through the adapter and button, and out as a stream, or spray. When the button is released, the spring forces the stem upward, creating a partial vacuum in the chamber and causing the ball to lift and allow product to flow upward to refill the body chamber with the product.

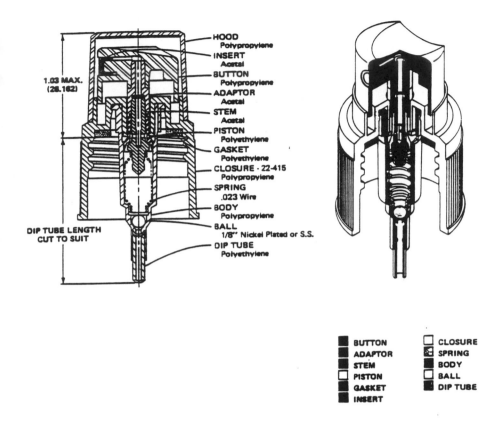

Figure 6.   Cross-Section of Finger-Action Seaquist Valve,
Set in 22-415 Closure

The complexity of the finger-pump valve systems makes them sensitive to strong solvents, solid suspension products and thixotropic viscous products. As a rule, they are only used for water-based, hydroalcoholic, and alcoholic formulations. The complexity of finger-pump valves is compared with the relative simplicity of a non-metering aerosol valve in the drawings presented in Figure 7. The complexity of the metering aerosol valve is more or less equivalent to that of the pump-action types.

Along with the complexity of these valves comes a considerable increase in cost, compared with aerosol valve options. Costs may be controlled by using the largest practical containers (to lower cost per unit of product), by marketing refill containers that use simple screw-caps, and by emphasizing the use of finger-pump valves with colognes, sachets, perfumes, pharmaceutical, and other generally high-cost products. In the case of perfumes and some medicinal items, the metering action of the finger-pumps is a distinct advantage in conserving and regulating the use of these products.

The pressure build-up within the chamber of the finger-pump valve is a complex function of the pressure applied to the actuator, the size of the exit orifices (diameter mostly, but also length), product viscosity, and other factors, but it is generally in the order of 50 psig (3.5 bars). Mechanical breakup spray heads do a fairly good job of developing spray patterns when the liquid is at a pressure of about 18 psig (1.27 bar) or higher. At excessive pressures (rarely attainable) there is some denigration of the pattern, such as "hot spotting."

The spray pattern and particle size distribution of finger-pump sprays is due to the purely mechanical breakup attributes of the specially designed two-piece button. Two to four tangential channels converge the product into a central swirl chamber, where it must turn at right angles to pass through the terminal orifice.

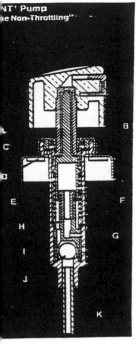

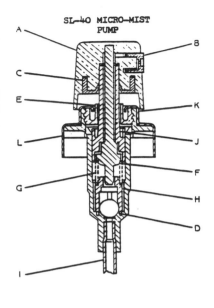

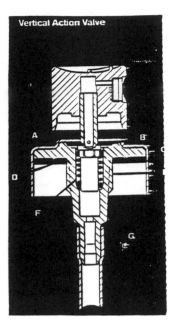

MATERIALS AND COMPONENTS IN VARIOUS VALVES

A	PISTON - Polyethylene	A	ACTUATOR - Linear Polyethylene	A	STEM - Nylon or Acetal
B	COLLAR - Polyethylene	B	ACTUATOR INSERT - Acetal	B	STEM GASKET - Buna-N
C	MOUNTING CUP - Alum.	C	ACTUATOR ADAPTER - Polypropylene	C	MOUNTING CUP - Aluminum
D	GASKET - Rubber	D	BALL - Stainless Steel #302 or #305	D	BODY GASKET - Buna-N
E	VALVE - Acetal	E	COLLAR - Linear Polyethylene	E	BODY - Nylon or Acetal
F	INNER PISTON - P.E.	F	BALL-SEAL INSERT - Acetal	F	SPRING - SS-302
G	SPRING - SS-302	G	SPRING - SS-302	G	DIP TUBE - Capillary PP with 0.020", 0.030" or 0.040" I.D., or 0.125" I.D. Polyethylene
H	BODY INSERT - P.E.	H	BODY & BODY INSERT - Polypropylene		
I	BODY - Polypropylene	I	DIP TUBE - Polypropylene		
J	BALL - Stainless Steel	J	PISTON - Linear Polyethylene		
K	DIP TUBE - Polyprop.	K	CLOSURE - Aluminum - anodized		
		L	GASKET - Rubber of Polyethylene		

Figure 7.   Comparison of Risdon (Dispensing Systems Division) Finger-Pump
20mm TNT Pump and SL-40 Micro-Mist vs. 20mm Aerosol Valve

Unlike most aerosols, where exploding actions caused by the instantaneous depressurization of liquified propellant act to reduce particle size to various degrees, the particle size of finger-pump sprays is regarded as very coarse, and best suited for surface applications. As a rule, spray particles from finger-pump units will strike the floor within five seconds or less, regardless of the initial direction of the spray. The only aerosols whose sprays compare with those of finger-pumps are "nitrosols," those pressurized with 1 to 6 grams of nitrogen gas (depending on size), and water-based types designed to have the hydrocarbon propellant separate on top as a discrete layer. Most of these latter products, such as starches and fabric finishes, carry a top-of-can message of "Shake Before Using" to obtain a better spray pattern by incorporating some propellant into the exiting product.

The finger-pump particle size distribution is compared with those of several similar products in Table 46.

Spray patterns of the finger-pumps vary from quite wide to very narrow. The valve of the "Moi•Stir" Mouth Moistener (Kingswood Laboratories, Inc.) will cast heavy droplets in a 7" (180-mm) slightly oval pattern onto a target panel 60" (1.51 m) distant. A "Hot Shot" Wasp & Hornet Spray (finger-action Calmar valve, ex. Bakan) by United Industries Corporation (St. Louis, MO), formerly Chemsico, Inc., will cast a narrow spray over 12 feet (4.35 m). Cologne sprays are usually the widest, with the particles traveling fairly slowly outward. In fact, cologne sprays, if deliberately ignited, will quickly burn back to the valve button and burn the fingertip of the operator, unless the spray shuts off first.

One of the detractions of the finger-action spray is the number of times the actuator must be depressed to empty the dispenser. For example, consider an 8.0-fluid ounce (299 mL) container, which is dispensed at the rate of 0.125 mL (125μL, or 100 mg) per shot. The required number of actuations to empty the dispenser will be 2,396. This number can be approximately halved by using finger-pump valves with 0.205 mL and similar size-metering dimensions. Many pump-spray marketers compensate for the slow use-up rate, compared with the

TABLE 46.   AEROSOL AND FINGER-PUMP HAIR SPRAYS: COMPARISON OF
PARTICLE SIZE DISTRIBUTIONS

		Particle Size Range (μ)[a]			
Type & Valve	% of Propellants	Below 10μ	10μ - 20μ	20μ - 50μ	Over 50μ
Finger-Pump mechanical breakup (MB)	0	0	2	14	86
Aerosol Non-MB	20	1	5	38	66
Aerosol MB	20	3	8	48	41
Aerosol Non-MB	25	2	8	49	41
Aerosol MB	25	5	15	60	20
Aerosol MB[b]	16.67	0.5	2	22.5	75
Aerosol MB[b]	32	16	18	39	27
Aerosol MB[b]	38	11	32	56	1
Aerosol Non-MB[b,c]	74	24	76	0	0

[a]Measurements made with a Malvern ST 1800 analyzer, at 90° to spray axis and
16" (406 mm) from the actuator. Run in duplicate.

[b]These are produced outside the U.S.

[c]The large amount of (CFC-11/12) propellant used in this product reflects the
high cost of ethanol in the country where it is produced; the "alcohol tax"
cannot be avoided, as in the U.S., for approved uses.

aerosol, by increasing the level of film-forming resin in the hair product formulation.

Some characteristics of pump-sprays are more economically attractive than aerosols. For example, a plain glass cologne bottle, is less costly than a heavier-walled, pressure-tested and PVC "Lamisol"-coated glass-in-plastic aerosol bottle. The plain bottle also has a number of other advantages relating to design flexibility. The filling operation for pumps is a single stage operation; the aerosols, however, must be filled and then gassed, requiring at least two stages. They must also be hot-tanked.

Aerosol containers larger than 4 fluid ounces (118.3 mL) are restricted to cylinders of aluminum or steel, at least in the U.S.; whereas, finger-pump dispensers may be made of various plastic or glass containers and be attractively shaped. Unlike aerosols, they are not limited to 819.4 mL in size, although very few are more than about 12 fluid ounces (355 mL), for practical reasons.

The flammability of aerosols and finger-pumps is commensurate in several ways. Formulas for both systems may range from 0% to 100% of flammable components. They pose approximately equal levels of hazard if exposed to an ongoing fire in a warehouse. The finger-pump can produce a flame volume of from 0.8 to 1.6 U.S. Gallons (3,000 to 6,000 mL) per actuation if the contents are hair spray or bug killer, which contain essentially 100% flammable ingredients. The aerosol is similar, but the flame volume may be two or three times larger and may be sustained by merely keeping the button depressed. Aerosols can rupture if overheated, and if a flame source is present, they may generate a fireball up to 9 feet (2.7 m) in diameter.

Typical products that have been successfully marketed in finger-pump sprayers include following:

Bug Killers (such as ant, roach, spider, and bee killers)
Weed Killers
Pet Sprays (often for insecticidal or grooming purposes)

Colognes and Perfumes

Hair Sprays

Hair Moisturizers

Curl Activators

Lens Cleaners (such as anti-fog and anti-static types)

Vermouth (for dry martinis)

Germicides (including those for pre-operation washing)

Spot Cleaners

Pre-suntanning Accelerator

Facial Rinse

Cookware Lubricant

Contact Lens Rinsing Sprays (requires Thimerisol or other disinfectant)

Window Cleaners

Topical Sprays (such as benzocaine or rubifacient types)

Silver Polish Sprays

Throat Sprays

Leaf Shine Sprays

Chrome Polishing Sprays (automotive uses)

Stainless Steel Cleaners

Mildewcides

One disadvantage of finger-sprays not yet discussed is that all models to
varying degrees produce extremely coarse dribbles at the very beginning and
the very end of each actuation. These heavy droplets fall downward very fast,
spotting polished wood furniture, window sills, flat glass surfaces and some
textiles, also cooling or wetting the skin away from the sprayed area of the
body.

Finger-pump sprays are usually not used with a number of product types
such as the following:

Volatile flammables   (such as cigarette lighter fluids)

Viscous liquids   (spray extra coarse, or may not spray)

Strong solvents   (such as nail polish removers & insect repel-
                         lents)

Sterile liquids   (sterility is lost at the first actuation)

Acidic liquids   (acetal valve components dissolve below pH =
                         3.6)

Moisture-sensitive   (moisture enters by return air and permeation)
liquids

Suspensoid fluids   (valve plugging can readily occur)

Foam-type emulsions   (foaming will not occur to any extent)

Polyethylene warping   (such as oleic acid or some block polymers)
liquids

Staining liquids   (such as food colors, dyes, etc. because of
                         dribble)

Sensitive liquids   (such as those harmed by air or light)

Two-phased liquids   (phases will reform in valve chamber and be
                         resistant to reconstitution by shaking)

High-odor liquids   (garlic concentrates, etc. will permeate)
(In plastic bottles)

In spite of all these apparent limitations, the finger-pump sprays enjoy a
business volume exceeding one billion units a year and remain the major
competitor to aerosols.

Trigger-Pump Sprayers

The sprayer is one form of the trigger-pump; the others extrude pastes, gels, or liquid products. Trigger-pump sprayers are more costly than finger-pump sprayers; they are used with somewhat larger dispensers, and more emphasis is given to refill units. The trigger mechanism facilitates the dispensing of larger quantities of product per shot, and the mechanical advantage or leverage feature of the pinioned trigger itself provides higher internal pressure in the chamber. They are generally viewed as more utilitarian than discretionary; for example, there are few if any trigger type pump-action hair sprays. (However, trigger pump lotions and cosmetic pastes are aesthetic and quite popular.)

Most trigger sprays are used for cleaning purposes, such as pre-laundry spot cleaners, disinfectant cleaners for hard surfaces, carpet cleaners, window cleaners, automotive cleaner and wax, vinyl top cleaners, industrial lubricant cleaners, and concrete floor (grease and oil) cleaners. Container sizes of up to one U.S. Gallon (3,784 mL) are available for institutional uses.

The operational principles, compatibility characteristics, and most other properties of the trigger-pump sprayers are equivalent to those of the smaller finger-pump sprayers and need not be repeated here.

Finger-Pump Extruders

A minor modification of the actuator changes the finger-pump sprayer into an extruder suitable for dispensing lotions, creams ointments, gels, pastes, viscous liquids, and measured amounts of various concentrates for dilution with fixed amounts of water. The actuator is removed and replaced with a spout with a very narrow tubular exit pipe. The small amount discharged per shot makes it useful for costly pharmaceutical, skin dewrinklers, perfumed lotions and similar products. In Europe, a vitamin mixture and an ear-wax softener are sold in this form. A concentrated cypermethrin and K-methrin mixture that is dripped onto an absorbent wafer measuring about 17 X 45 X 2 mm

in size is also sold in this form. The treated wafer is slipped into a small
holder that plugs into a wall socket which gently heats it to vaporize the
insecticidal additives. The active ingredients are not volatilized in
sufficient concentrations to be lethal, but they are so irritating to mosquitoes that they vacate the room if possible. The repellent action lasts 8
to 10 hours, ensuring people a good night's sleep. Most of the sales are in
Mediterranean countries, where the product has made serious inroads into the
much more costly aerosol insecticide business.

Trigger-Pump Extruders

Various modifications can be made to the trigger-pump sprayer to change
it to a device with a spout able to dispense lotions, gels, and similar
products in the form of a stream or ribbon. Simplified and lower-cost
versions are also in demand that are used to discharge relatively large, fixed
volumes of dishwashing detergents, fabric softeners, and other cleaners. They
will have almost no effect on the aerosol market as possible alternatives and
are not discussed further.

DISPENSING CLOSURES

One of the simplest possible designs is the screw-threaded closure or cap
with a dispensing hole able to be plugged shut by various means. Three of
these designs are illustrated in Figure 8.

To operate these, the dispenser is held inverted to get the product near
the orifice, after which, the "F"-style metal can (oblong, with large front
and back flat surfaces) or flexible plastic container is squeezed, expelling a
stream or ribbon of the product. Dispensers come in sizes of 6 to 64 fluid
ounces (177 to 1,892 mL) and can conveniently dispense liquids, thin gels,
soft creams, and lotions, as long as they are flowable. These containers are
used for charcoal lighters, various cosmetics, toiletries, personal care
products, paint solvents, paint strippers and furniture polishes.

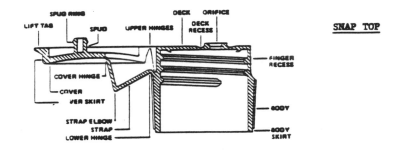

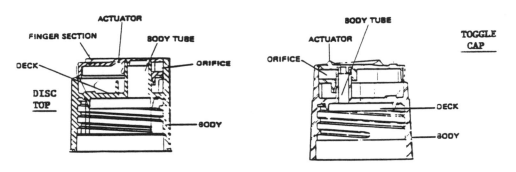

Figure 8.   Various Dispensing Closures-Made by the Seaquist Closures Division

In addition to the designs illustrated above, there are turret spouts
(truncated cone profile), lever spouts--in which a small pinioned plastic
section is rotated 90° upward to operate the closure--and several related
forms. They are increasingly used instead of the simple, detachable screw-cap
dispensers. Uses include certain foods (such as oleomargarine pourables,
ketchup, and mustard), lubricants, silicone shoe and boot dressing, some
medicinals, and solvents for home use, artists, and industry.

These products are major competitors with aerosols in the lubricant field
(aerosol volume 95,000,000 units in 1988), for carburetor and choke cleaners
(aerosols 57,000,000), waxes and polishes (aerosols 129,000,000), and certain
other overlap product areas. Since the closure is a single polyethylene
molded unit, generally applied semi-automatically as a replacement for screw-
threads, it is a very economical option. Some models can be made child
resistant.

PRESSURIZING DISPENSERS

Twist-N-Mist II

Over the years, several firms have developed various pressurized packaging systems quite different from the conventional aerosol form. They invar-
iably use air pressure, the restorative pressure from an expanded rubber
bladder, or some similar arrangement as the dispensing method. They are
characterized by delivering either very coarse sprays or various lotions and
semi-solid products, usually one or the other.

The Twist-N-Mist II is a development of the CIDCO Group, Inc. of Denver,
CO, which holds several U.S. Patents that cover the principles of the device,
as well as those employed in related dispensers: Pull-N-Mist and Dial-A-
Spray, details of which are still experimental and have not yet been released.
The firm also holds several foreign patents.

As in all such products, energy must be imparted to the dispenser to take
the place of the propellant gases used in aerosol forms. For Twist-N-Mist II,

that energy is supplied manually, by rotating the full-diameter screw-cap and integral piston.

The current model of Twist-N-Mist II, of which several hundred have been made, uses a three-component outer shell assembly, which measures about 2 3/4" X 6 1/2" (70 X 165 mm) and consists of an HDPE threaded base, threaded top, and matching body, as shown in Figure 9.

By turning (twisting) the threaded cap several revolutions the integral piston in the base of the cap is raised about 1/2 inch (12.7 mm) or so, creating a vacuum in the cylinder (reservoir) below. This causes the product to rise up the dip tube, past the stainless steel ball check valve, to fill the cavity. Enough is drawn up to provide a 7- to 20-second spray time, depending on the valve orifice.

The cap is now twisted an equal number of turns in the opposite direction, forcing the integral piston downward until it hits against the base of the reservoir. This action causes pressure to develop in the reservoir and forces the trapped product downward into a Buna-N rubber bladder, which expands accordingly. The memory of the elastomer causes pressure, which decreases to some extent as the product is dispensed through an aerosol type valve, allowing the bladder to slowly regain its original "test-tube-like" profile. The process must be repeated for another actuation. The dimensional changes in the unit during the suction and pressurization stages are shown in Figure 10.

As a fail-safe feature, the contents of the pressurized rubber bladder will very slowly bleed back past the check valve barrier and into the main product storage compartment. The bleed-back time can be controlled by varying the surface finish of the check ball or check ball receptacle, or, if the product is viscous, by grooving the ball.

A number of other features are possible. The amount of pressurized product (and thus spray time) can be pre-engineering by proper sizing of the reservoir, bladder and/or nozzle orifice. The main section of the dispenser,

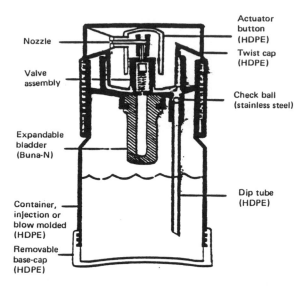

Figure 9.   Twist-N-Mist II

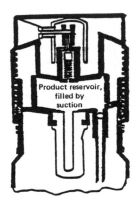

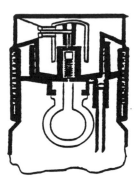

FILLING RESERVOIR   Turning cap to the up position opens the reservoir and fills it through suction on the dip tube.

DISPENSING SEQUENCE   Twisting cap back to the down position forces the product from reservoir to bladder. Pressing actuator discharges contents.

Figure 10.   Suction and Pressurization Stages of the Twist-N-Mist II Dispenser

made from injecting blow-molded HDPE or HDPP, can be contoured to a modest degree inward, outward, or both, if the screw-threaded top and bottom sections remain round. Technically, the dispenser can be provided with an integral bottom at a slight decrease in cost, but this would make it into a one-time service unit, instead of a reusable type and increase cost-per-ounce (cost per mL) significantly. This option is not generally recommended.

Because the upper cavities are completely filled with product, the unit may be used with the dispenser held in any direction. It delivers about 95+% of the contents. Corrosion is not a problem, since the only metal parts are a stainless steel spring and ball. Stress cracking has been noted as a problem with early single cavity models, mainly affecting the screw-threaded dome section and allowing leakage of the product. If refined models are resistant to stress cracking, they should be tested with surfactant (as non-ionic) water solutions that can often induce this problem in polyethylenes that are not formulated properly.

The CIDCO Group, Inc. recites the shortcomings of aerosols (their major target) as well as of finger-pumps and trigger-pumps, claiming that their dispenser, while somewhat costly to buy the first time, has long-range advantages, especially if refilled.

However, several turns may be necessary to pull product from the main chamber and then force it into the Buna-N Rubber bladder against the back pressure from that diaphragm. The spray duration could cease in the middle of a spray episode, requiring the user to delay completion for an estimated 15 to 30 seconds to recharge the can. The spray is much coarser than that of aerosol sprays, except for nitrosols. No foams can be produced. No solvents that have a profound swelling or deleterious effect upon the Buna-N bladder can be used, except perhaps at low concentrations. Product darkness or odors may develop unless the rubber bladder is specially lined, as in the Exxel system discussed below.

The Exxel System

Briefly, this is another dispenser option where the product is contained in a thick, squeezable rubber sleeve open at one end, but in this case all the product is compressed into the inner container by a manufacturer or packager. The dispenser has aerosol properties, in that the product is always under pressure. However, there are differences. Sprays are propellant-free, and therefore very coarse or wet, and no foam type products can be provided except for post-foaming gel types.

The following steps are required to manufacture the Exxel System dispenser:

- Stretch-blow a biaxially oriented, thin-walled polyethylene teraphthalate (PET) bottle;

- Form longitudinal pleats in the bottle, using patented equipment;

- Apply a double layer of barrier sealant and liquid latex to the bottle;

- Insert a customized valve and clinch in place at the top ring of the bottle;

- Insert bottle into a rubber sleeve;

- Place container into a suitable outer container and attach at the top; and

- Force a predetermined amount of product into inner container via the valve.

Construction materials that can contact products are limited to the PET bottle, the Nylon 66 valve housing, natural polypropylene, the HDPE button, and either the SS #302 or #316 spring. An insignificant exposure to the PET/valve gasket must be mentioned. The gasket is available in various materials.

A sampling of products currently being packed in the Exxel system appears in Table 47.

Exxel comments that skin care, hair care, and pharmaceutical products of the post-forming gel type are well along in the development stages. Also, several medicinal ointments are under intensive study by Upjohn and others. Cost comparisons can be made with other forms of packaging, using the tabulated data in Table 48.

The Exxel system is incompatible, to varying degrees, with certain formulations. Following is a list of ingredients and characteristics that would make a product incompatible with the Exxel system:

- Certain polymer solvents--terpenes, ketones, etc.;

- pH Values over 10.0;

- Isopropanol, above 5.0%;

- Prolonged exposures to over 113°F (45°C);

- Particulate matter--since effective shake-before-use is impossible;

- High surface tension breakup products;

- Resins with an ability to dry and clog actuators;

TABLE 47.   TYPICAL CURRENT CUSTOMERS AND PRODUCTS OF THE EXXEL SYSTEM

Company	Product
Air Products and Chemicals Company	Welding Flux Spray
Chanel, Inc.	Sun Oil Spary
Kobayashi Pharmaceuticals Company	Muscle Relaxant
Nihon Sanso, Ltd.	Pure Food Products - sterile
Prudue Frederich, Inc.	"Betadyne" Solution[a]
P.R. Hertensen	"Citruscent" Fragrance
Tokyo Aerosol Co.	Hair Gel
Wella	Shampoo and Conditioner
Jergens	Topical Lotions[b]
Westwood Pharmaceuticals, Inc.	"Alpha Keri" Spray Oil
Adrien Arpel	"Aromafleur" Flower Extract Foam Firming Masque
Estee Lauder, Ltd.	Hair Reviving Mist
HiLo Products	"Silent Force" Flea Spray
Laboratoires Goëmar, S.A. (France)	"Tonialg" Restorative Conditioner "Tonialg" Toning Lotion "Tonialg" Night Creme "Tonialg" Restorative Shampoo "Tonialg" Hand & Body Creme "Tonialg" Bath & Shower Gel "Tonialg" Foaming Bath "Tonialg" Cleanser "Tonialg" Nourishing Creme "Tonialg" Body Contouring Creme

[a]Tamed Iodine formulation.

[b]As of 1988.

TABLE 48.   COST OF EXXEL SYSTEM PACKAGING AND FILLING SERVICES
(VOLUME — 1MM)

Item	4 oz. ($/M)	7 oz. ($/M)
Snap Ring	16	16
Power Assembly Unit	241	274
Actuator	30	30
Overcap	22	22
Decorated Bottle	85	95
Filling Charge (Contract)	140	140
	534	577

NOTE: Add $10M for 500,000 quantities and $20M for 250,000 quantities.

Add $25/M for pre-fill electron beam sterilization.

TABLE 49.   MINIMUM DIMENSIONS OF OUTER CONTAINERS FOR EXXEL UNITS

Dimension	4 oz. Size	7 oz. Size
Minimum length. Top of neck finish to inside of bottom.	5.450 in. (138 mm)	7.300 in. (185 mm)
Minimum width. Internal dimension.	2.223 in. (56.5 mm)	2.223 in. (56.5 mm)

Note: Intentionally underfilled Exxel units will permit the use of outer containers with reduced minimum internal widths.

Outer containers require a vent hole of at least 0.015" (0.38 mm), preferably in the base, but alternatively in the shoulder.

- Formulas that require in-package mixing, e.g., bi-phasics; and

- Ethanol, above 60%--should be carefully tested for compatibility.

The outer container may be made from glass, plastic metal, composites, paper, or (theoretically) nothing at all. For automated filling, the container should be able to support a hold-down filling force of 25 pounds (11.4 kg) without buckling. To accommodate the two Exxel System inner containers, the minimum internal dimensions of the outer containers must be considered, as shown in Table 49.

The smaller (4-fl.oz.) Exxel package will deliver 92 to 95% of its contents before fully depressurizing. The 7-fl.oz. size will deliver 92 to 84.7%. These ranges are reduced to 90 to 93% in the case of fairly viscous items with positive yield points. After about 86 to 88% of spray products have been dispensed, the bag pressure falls below about 17.5 psig (1.23 bar) and the spray pattern deteriorates rapidly. The pressure at which this occurs depends on the ingredients and the viscosity of the formulation. As a rule, storage weight loss will be 1.0% per year at ambient temperature or per month at 104°F (40°C).

The Exxel System is self-pressurized and may be classed as an aerosol under the DOT shipping regulations; however, DOT Exemption No. E-9607 has been obtained by the Darworth Company (Div. of Ensign-Bickford, Inc.) to set aside these requirements for hot tanking, etc. This now applies to all Exxel System products.

The Mistlon System

The Mistlon Eco-Logical Spray Bottle is a dispenser developed in Japan, made in South Korea, and offered for sale by the MONDEX Trade & Development Corporation, 2 St. Clair Avenue - West (Suite 801), Toronto M4V 1L5, Canada. It is cylindrical and measures 2 1/8" X 8 1/2" high (54 X 216 mm). The wholesale price is about $1.00.

To use the empty unit, the full-diameter polypropylene cap is removed, after which a screw-threaded closure carrying an ordinary aerosol valve and actuator is also removed. A quantity of product is poured into the bottle through the one-inch (25.4 mm) opening. A typical fill is 250 mL. After this, the closure is screwed back into place, allowing a thin rubber "O"-ring to make a reliable hermetic seal. The base section [full-diameter and 1" high (25.4 mm)] is now withdrawn, away from the rest of the unit, exposing a hollow cylinder 11/16" in diameter by 3 11/16" long (17.5 X 93.7 mm), fitted with a one-way compound valve. The hollow cylinder functions as a piston, within a cylinder protruding upward into the container, also ending in a one-way valve. To pressurize the air in the container, the base section is pumped a number of times. By pressing a soft diaphragm in the center of the base, excess air pressure within the hollow cylinder is removed, allowing it to fit snugly against the bottom-most area of the body as before.

The unit is equipped with either a 0.010" (0.25 mm) or 0.014" (0.36 mm) mechanical breakup bottom. In the case of water, these actuators will provide an acceptable spray if the air pressure is 18 psig (1.17 bar) or greater. The operating characteristics follow simple gas laws. This can be illustrated by the following example.

Characteristics:

> The head space volume is 100 mL.
> The liquid volume is immaterial.
> The applied pressure is 50 psig (3.52 bar).

Question:

> How much product can be dispensed before the pressure sinks to 18 psig (1.17 bar) and the spray starts to deteriorate?

Solution:

Convert to absolute pressures.
        50 psig - 64.7 psi-abs (4.56 bars - absolute)
        18 psig - 32.7 psi-abs (2.30 bars - absolute)

Boyle's Law:

V_2 - V (P /P_2) - 100 mL X (64.7/32.7)
V_2 - 197.9 mL

Change in head space volume.

V_o - V_2 - V - 197.9 - 100.0 - 97.9 mL.

Answer:

97.9 mL of liquid can be dispensed before the spray deteriorates.

It follows that, the larger the head space, the more strokes of the piston will be needed to pressurize to a given level, and the more liquid can be dispensed as a result.

With some degree of manual difficulty, the Mistlon unit can be pressurized to 65 psig (4.58 bars). There was no evidence of deformation at this pressure. The unit might be pressurizable to well over 100 psig (7.04 bars) without any problems unless it is strongly heated to the point where the polypropylene begins to soften and become deformable.

The delivery rate will, of course, vary with the container pressure. With the 0.014" (0.36-mm) MB valve button, water delivers at about 0.5 g/s at 20 psig (1.41 bar) and about 0.72 g/s at 40 psig (2.82 bar).

Like the Twist-N-Mist dispenser, the unit is limited in terms of spray particle size and range of products that can be dispensed. Highly flammable materials, those that deform polypropylene or attack polyvinylacetate, and viscous fluids are among those that should not be used. The unit cannot be used to generate a direct foam, but with a suitable straight bore actuator it could direct a thin stream into the palm that would then spring into a foam.

Airspray

This product, which is similar to the Mistlon dispenser also uses a pumping action to compress air into the pressure-resistant container. When reasonably full, it must be pumped 10 to 20 times to create an effective spray that does not quickly deteriorate as pressure drops. As the container empties, the number of pumping strokes must be increased, but the pressure lasts longer during dispensing. The unit must be held upright to keep the diptube below the liquid surface, but if the container is held so that the compressed air is unloaded, it can be quickly pumped up again, unlike aerosol products with nitrogen or other propellants in low concentrations.

The cylinder is cylindrical to withstand the generated air pressures without buckling or other deformations. Comments made about the Mistlon dispenser apply here as well.

Invented in Sweden, the Airspray system was developed and refined by a Dutch company, which marketed the unit in Europe for several years. In 1987 they entered into an agreement with the National Can Corporation, which is now a part of the American National Can Company unit of Pechiney, S.A., to manufacture and market the system in the U.S. under license. As of 1989, the system will be jointly marketed in the U.S. by Airspray International, Inc. (Pompano Beach, FL) and American National Can Company (Chicago, IL). It is promoted in Canada by W. Braun & Company (Markham, Ontario L3R 3B3, Canada).

The system is offered in two versions: with a refillable screw top and a disposable crimp-on. It can be made in containers of plastic, metal, or glass. PET containers are being developed at this time. All the parts are plastic. Once pressurized to 55 psig (3.87 bars)--the recommended maximum--it will dispense up to 100 mL before repumping is needed. Airspray supplies an

O.T.I. crimping machine for closing and pressurizing the system at 1500 μ/hr with compressed air.

The Werding Nature Spray-Systems

A variety of inter-related systems have been developed by Werdi Spray, S.A. 5, Route des Jeunes, CH-1227 Geneva (Switzerland). They are represented in the U.S. by Werding Aerosol Technology Inc., U.S., located at 4978 Kingsway, Burnaby, British Columbia V5H 2E4, Canada.

The firm makes both non-aerosol containers and their unique Werdi 'R' Actuator. The latter can be designed to provide a constant delivery rate, regardless of the internal pressure of the dispenser, and is thus most useful for products pressurized with air, nitrogen, carbon dioxide, etc., where pressure drops during use can exceed 70%. The Werdi 'R' System comprises the Werdi 'R' Actuator (fitted with the Werdi 'N' Nozzle and thrust regulator) and the Werdi Valve. For lotions and creams, the Werdi 'RD' system is suggested, which consists of the Werdi 'RD' Actuator (fitted with the thrust regulator and a self-closing diffusor) and the Werdi Valve.

The Werdi 'N' Nozzle achieves a high mechanical breakup effect by means of its multi-staged, interconnected Venturi system, and thus contributes more to spray breakup than conventional (less costly) mechanical breakup actuators. Fitted behind the nozzle in the actuator, the thrust regulator controls the flow of product to the nozzle. The patented design includes two stainless steel accelerator discs and a plastic expansion chamber as well as a special regulation disc, which is cut, curved, and formed to exacting standards.

The regulation disc is compressed by higher pressures, but because of the spring effect of the metal, this opens the cut and increases the orifice size as the pressure drops. Turbulence intentionally created by the design of the companion discs, as well as the nozzle itself, produces a resistance to the product flow into the thrust regulator, whose force is directly proportional to the pressure. The higher the pressure, the more these turbulent effects brake the delivery rate. Thus, the Werdi 'R' Actuator maintains a constant outflow of product from the container.

There are four types of the Werdi 'N' Nozzle, which are used for different spray rates and patterns. When a nonaerosol (or aerosol) container is filled to 65 volume percent with low-viscosity concentrate and then pressurized with air or nitrogen to 85 psig (6 bars), the results are as shown in Table 50.

Werdi also makes complete valves as well as nonaerosol (pump-type) containers, but their primary contribution to nonaerosol dispenser technology appears to be in the actuator area. The following U.S. Patents are reference sources: 4,487,554 (11-DEC-84), 4,260,110 (7-APR-81), and Battelle's 4,603,794 (5 AUG-86). The last describes a dispenser able to deliver a high-pressure spray by means of a low-pressure squeeze on the flexible sidewall area, following a pressure multiplying principle.

Latest reports suggest that a large Northern Italian watchmaking firm is interested in purchasing Werdi because they have facilities to produce many of the very small actuator and other parts required for the system.

MISCELLANEOUS AEROSOL ALTERNATIVES

A number of dispensers can be used to present products that compete with the aerosol system, although they may bear no direct similarity to aerosols. Two will be considered in the following pages.

Insecticide Vaporizers

Vaporizers of various types have been used to provide "true aerosol" mists or condensation nuclei of products in the air. For the most part, they have been used for insecticides, but triethylene glycol mists of hexylresorcinol and other health-related products have enjoyed a much smaller market.

TABLE 50.   SPECIFICATIONS--USING FOUR NOZZLES--FOR THE WERDI 'R' ACTUATOR

Nozzle	Type A	Type B	Type C	Type D
Color Code	White	Yellow	Green	Black
Average Delivery Rate (mL/sec.)	0.70	0.47	1.30	1.30
Average particle Size (microns)	3[a]	5[a]	0.7 - 3[a]	25 - 50
Cone Angle of Spray Pattern	50°	30°	40°	30°
Cone Length (inches)	30	24	55	36
Range of Applications	Personal Deodorants	Hair Spray	Space Insecticide	Polishes
	Pre-shaves	Wound Spray	Air Fresheners	Surface Insecticide
	Leaf Polish			
				Surface Disinfectants
	Mold Releases			

[a]The average particle size for Types A, B, and C appear to be unusually low for air sprays.

In Latin America, Spain, Portugal, Tripoli, and other areas the electrically vaporized insecticide products form the largest single use for insecticide applications. Individual insecticide wafer sales volumes are greater
than the total aerosol markets in these countries. Such well-known firms as
S.C. Johnson & Son, Inc., Refinacoes de Milho, Brasil, Ltda (STP Brands),
Bayer, GmbH (BAYGON Brands), and Reckett & Coleman, Ltd. (Various Brands) sell
the wafers.

The wafer, which contains a few drops of absorbed insecticide concentrate, is placed in a holder on the heater, which is then connected to a
wall plug of electric current. The wafer is gently warmed to release the
insecticide materials. Although nontoxic at low levels of use, the insecticides irritate mosquitoes (and "permilongos"--long-legged mosquitoes) so that
they leave the room. Especially useful in sleeping quarters, the wafer has
useful service life of from eight to ten hours. Foil packs of these products
are now being replaced with PET-laminate packs to reduce packaging costs.

Stick Products

Coming into major use only about ten years ago, the stick-in-canister
option has become the leading alternative for antiperspirants and personal
deodorants. A much smaller market exists for other items such as stick insect
repellents, stick spot-cleaners for textiles, stick analgesics (methyl
salicylate types, for example), and several other products.

Several types of polyethylene and polypropylene round and oval canisters
exist. The most popular are in the 1.5- to 3.5- Av.oz. (42.5 to 99.2 g) size,
with a bottom-entering plastic screw that, when rotated, elevates the product
so that it protrudes sufficiently from the top of the canister to allow for
convenient use.

A typical stick antiperspirant formulation contains 20 to 25% of the
aluminum chlorohydrate complex salt, compared with 7 to 12.5% in aerosol
products. Two representative formulas appear in Table 51.

TABLE 51.   TWO STICK ANTIPERSPIRANT FORMULAS

Antiperspirant Stick		Improved Antiperspirant Stick Formula	
Ingredient/CTFA Name	%	Ingredient/CTFA Name	%
Phase A		(A) Cyclimethicone	43.5
Permethyl 99A[a] Isododecane	17.15		
Permethyl 101A[a] Isohexadecane	4.00	Stearyl Alcohol	23.0
Dow Corning 244[b] Cyclo-		PPG-15 Stearyl Ether	
methicone	13.15	(ARLAMOL E)	5.0
Fluid A/P[c]PPG-14 Butyl Ether	11.50	(ICI Specialty Chemicals)	
Phase B		(B) Hydrogenated Castor Oil	2.0
Crodacol S-95[c] Stearyl			
Alcohol	11.50	Steareth-20 (BRIJ 78)	1.0
Castorwax MP-80[d] Hydrogenated		(ICI Specialty Chemicals)	
Castor Oil	7.50		
		(C) Silica	0.5
Phase C			
Micro-Ace P-2[a] Talc	10.50	Aluminum Chlorohydrate	25.0

Phase D
Spheron P-1500[a] Silica   2.00

Phase E
Micro Dry[e] Aluminum Chloro-
    hydrate   22.00

Manufacturing Procedure: Add Phase
A in order to vessel, heat to 70-
75°C. Mix until clear and uniform.
While mixing, add Phase B one item
at a time. Continue mixing until
clear and uniform. Maintain 70 to
75°C, add Phase C, keep agitation
vigorous. Add Phase D, mix for 5-
10 minutes. Pour into containers
at 66-68°C.

Procedure: Heat (B) to approximately
65°C until liquid. Add (A) with
moderate agitation and heat to minimize
silicone evaporation. Add (C) and stir
5-10 minutes until uniform. Cool to
55°C with stirring and pour into stick
forms.

Suppliers:
    [a]Presperse Inc.
    [b]Dow Corning
    [c]Croda
    [d]Cas Chem
    [e]Reheis

All significant marketers of aerosol underarm products also sell the
stick products. Each line generally has two sizes and "scented" and
"unscented" versions. Product effectiveness is equivalent to, or somewhat
higher than, those of the latest generation of aerosols, and the "antiperspirancy" of both versions is well above FDA requirements.

The antiperspirant type of underarm product commands 81% of the total
underarm aerosol business, and 83% of the stick alternative. The personal
deodorant subsegment is presented in the same container types and sizes.
Instead of aluminum astringent salt, it contains 0.1 to 0.2% of a germicidal
material, typically Triclosan, a diphenyl derivative made by Ciba-Geigy
Corporation. Table 52 shows approximate production volumes of aerosol and
stick underarm products.

Other packaging forms, including roll-ons and pads, make up a relatively
minor proportion of the U.S. market. These secondary alternates will not be
covered here.

Aerosol and stick underarm products are mature markets. The change in
ratio shown in Table 52 is the result of new antiperspirant entrants (Bristol-
Myers and Mennen) whose advertising helped both their products and the aerosol
packaging concept. In addition, reformulation to more powerful forms of the
aluminum chlorohydrate have made aerosol antiperspirants more effective.
Unless significant changes in price structure, ecological aspects, flammability considerations, or other criteria affect one product at the expense
of the other, the 1:1.50 ratio of aerosols to sticks will probably continue
for a long time. No dramatic changes are seen in this ratio for at least four
years.

TABLE 52. PRODUCTION UNITS OF UNDERARM PRODUCTS (U.S.)

Year	Aerosols	Sticks	Ratio
1986	153,000,000	258,000,000	1:1.69
1987	164,500,000	278,000,000	1:1.69
1988	193,000,000	292,000,000	1:1.51
1989[a]	207,000,000	310,000,000	1:1.49

[a]Estimated figures at mid-1989.

3. Summary

Part I of this report discusses the aerosol industry's experience in converting from CFC propellants to alternative aerosol formulations. Some of the immediately available alternatives, such as HCFC-22 and 1,1,1-trichloroethane, also can deplete stratospheric ozone levels, although their ozone depletion potentials are less than those of the CFC propellants.

Such compounds as HCFC-123, HCFC-124, HFC-125, HCFC-132b, HCFC-133a, HFC-134a, and HCFC-141b are now undergoing extensive toxicological testing that will continue until about 1992. Many of these "future alternative" compounds are nonflammable unless they are mixed with substances such as iso-butane or ethanol; others are flammable. Hydrocarbon propellants, which cost less than CFCs, are often the propellants of choice unless special properties such as increased solvency or reduced flammability are needed. Dimethyl ether (DME) is the next most preferred CFC alternative. DME is flammable and a strong solvent.

Carbon dioxide, nitrous oxide, and nitrogen are inexpensive and widely available throughout the world but have been underused as aerosol propellants. Special equipment is often needed to add them to the aerosol containers.

As CFC suppliers in the U.S., Western Europe, Japan, and other parts of the world develop their CFC phase-down programs, which will go beyond the Montreal Protocol, they will be focussing on rapid commercialization and application of the HCFC and HFC alternatives. The major alternative will be HFC-134a, which will replace CFC-12 in refrigeration, freezant, and air conditioning systems.

A variety of alternative aerosol packaging forms has been discussed in Part II. with a special focus on those most like regular aerosols in characteristics. All the alternatives have subsidiary positions in the marketplace, if the volume of each is compared with the 3,000,000,000-unit volume of aerosols. Several have been available for many years but have not significantly penetrated the market for several reasons, shown below:

> They generally cost more (finger-pumps and sticks are exceptions).

> They are limited in their product compatibility.

> They depend on chemical or mechanical (often manual) action to generate pressures needed to discharge the contents.

> Products must be delivered as very coarse streams. pourables, paste ribbons or (sometimes) post-foaming gels -- without having the broad range of the aerosol presentation.

> Sterility is generally impossible.

> Sprays can deteriorate during use.

> Several are incompletely tested.

> Several require capital expenditures for special filling or gassing equipment.

> Sizes are limited to the 3-fl.oz. to 12-fl.oz. (119- to 355-mL) range (some are even more limited).

In general, the packaging alternatives continue to be niche-fillers, working best for a very limited range of products. Sales volumes are expected to grow to some extent, however, taking some market share away from aerosols **in selected areas, but without significantly affecting the aerosol** business if **the present mix of political, regulatory, economic, environmental, financial, and other issues remains reasonably static.**

Appendix A—Metric (SI) Conversion Factors

Quantity	To Convert Form	To	Multiply By
Length:	in	cm	2.54
	ft	m	0.3048
Area:	in^2	cm^2	6.4516
	ft^2	m^2	0.0929
Volume:	in^3	cm^3	16.39
	ft^3	m^3	0.0283
	gal	m^3	0.0038
Mass (weight):	lb	kg	0.4536
	oz	kg	0.0283
	short ton (ton)	Mg	0.9072
	short ton (ton)	metric ton (t)	0.9072
Pressure:	atm	kPa	101.3
	mm Hg	kPa	0.133
	psig	kPa	6.895
	psig	kPa*	((psig)+14.696)x(6.895)
Temperature:	°F	°C*	(5/9)x(°F-32)
	°C	K*	°C+273.15
Caloric Value:	Btu/lb	kJ/kg	2.326
Enthalpy:	Btu/lbmol	kJ/kgmol	2.326
	kcal/gmol	kJ/kgmol	4.184
Specific-Heat Capacity:	Btu/lb-°F	kJ/kg-°C	4.1868
Density:	lb/ft^3	kg/m^3	16.02
	lb/gal	kg/m^3	119.8
Concentration:	oz/gal	kg/m^3	
	quarts/gal	cm^3/m^3	25.000
Flowrate:	gal/min	m^3/min	0.0038
	gal/day	m^3/day	0.0038
	ft^3/min	m^3/min	0.0283
Velocity:	ft/min	m/min	0.3048
Viscosity:	centipoise (CP)	Pa-s (kg/m-s)	0.001

*Calculate as indicated

Other Noyes Publications

OSHA REGULATED HAZARDOUS SUBSTANCES

Health, Toxicity, Economic and Technological Data

Occupational Safety and Health Administration
U.S. Department of Labor

This two-volume book provides industrial exposure data and control technologies for more than 650 substances currently regulated, or candidates for regulation, by the Occupational Safety and Health Administration (OSHA). The health, toxicity, economic and technological data provided for each substance are intended to serve as a reference for those who are potentially exposed to one or more of these substances in their workplace, or for those who have supervisory or management responsibility for workers potentially exposed. OSHA "permissible exposure limits" (PELs) for these 650 substances reflect all updates and changes as presented in the *Federal Register*.

The information on each substance in the book includes, as available, synonyms, trade names, physical description, health effects, toxicity/exposure limits, industry use data, and NIOSH National Occupational Exposure Survey data, NIOSH National Occupational Hazard Survey data, OSHA/exposure data, engineering controls, personal protective equipment, and storage.

Indexes included in the book provide cross referencing by synonyms, trade name, and Chemical Abstract Service (CAS) number.

Below is an alphabetical listing of the first 61 substances in the book.

Acetaldehyde
Acetic acid
Acetic anhydride
Acetone
Acetonitrile
2-Acetylaminofluorine
Acetylene
Acetylene tetrabromide
Acetylsalicylic acid
Acrolein
Acrylamide
Acrylic acid
Acrylonitrile
Aldrin
Allyl alcohol
Allyl chloride

Allyl glycidyl ether
Allyl propyl disulfide
alpha-Alumina
Aluminum: alkyls
Aluminum: metal and oxide
Aluminum: pyro powders
Aluminum: soluble salts
Aluminum: welding fumes
4-Aminodiphenyl
2-Aminopyridine
Amitrole
Ammonia
Ammonium chloride (fume)
Ammonium perfluorooctanoate
Ammonium sulfamate
Amosite
n-Amyl acetate
sec-Amyl acetate
Aniline and homologues
Anisidine (o-, p- isomers)
Antimony and compounds
Antimony trioxide as Sb
ANTU
Argon
Arsenic and compounds, as As
Arsenic trioxide production
Arsine
Asbestos
Asphalt fumes
Atrazine
Azinphos-methyl
Barium sulfate
Barium, soluble compounds
Benomyl
Benzene
Benzidine
Benzidine (based dyes)
Benzo(a)pyrene
Benzoyl peroxide
Benzyl chloride
Beryllium and compounds
Biphenyl
Bismuth telluride
Borates, tetra, sodium salts
Boron oxide
            plus 589 other substances

ISBN 0-8155-1240-6 (1990)   6" x 9"   2 volumes   2294 pages

Other Noyes Publications

COUNTERMEASURES TO
AIRBORNE HAZARDOUS CHEMICALS

by

J.M. Holmes and C.H. Byers
Oak Ridge National Laboratory

Pollution Technology Review No. 182

Recent major incidents involving the release of hazardous chemicals have heightened the awareness of both the public and the private sectors that effective strategies must be developed to prevent and to deal with emergencies. A number of federal, state, and local government agencies share portions of the responsibility for various aspects of the problem. This book presents a study which reviews the entire spectrum of activities, recommends appropriate action, and gives technical guidance.

Examples of technical information included are:

    Foam Systems for Vapor Suppression
    Commercial Foams
    Applying Vapor Suppression Foams
    Chlorine Emergency Kits
    Inert-Gas Systems
    Handling Spilled Fuels and Chemicals
    Steam Smothering Systems
    Combined-Agent Systems
    Protective Clothing
    Clothing Selection
    General Eye Protection
    Respiratory Protective Equipment
    Emergency Warning Systems
    Depressurizing
    Secondary Containment Systems
    Reduction of Toxic Material Inventories
    Substitutes for Hazardous Materials
    Explosion Suppression Systems
    Remotely Operated Response Equipment
    Advanced Computer Systems
    Controlled Burning of Hazardous
      Substance Releases

A condensed contents listing **chapter titles and selected subtitles** is given below.

1. **INTRODUCTION**

2. **OVERVIEW OF RECENT CHEMICAL EMERGENCIES**

3. **RELATIVE ACCIDENT FREQUENCIES AND SEVERITIES**

ISBN 0-8155-1232-5 (1990)   6" x 9"   330 pages